TAP & LET GO

life after trauma

TAP & LET GO

life after trauma

Jeannette van Uffelen

YIP
BOOKS

TAP & LET GO – life after trauma
Written by Jeannette van Uffelen

First edition: April 2024
Copyright © 2024 Jeannette van Uffelen

Published in Dutch in October 2022:
LEKKER LOS LATEN – leven na trauma,
Copyright © 2022 Jeannette van Uffelen

Translated by Dominique Pickerill

Cover photo by CJ Kale

Published by YIP Books
ISBN 979-82-24199-47-1

What people say about this book

An intelligent book and a masterpiece. Cases as a process around the theme of 'tap and let go', which runs like a common thread through the author's own life. She now offers the readers to enter into that process in an easy and relaxing way to heal their trauma. Deep bow.

Lynn Hogendoorn, stressologist on sneakers

A beautiful and inspiring book. An easy read! Some parts really touched me enormously and have made clear some of this for myself again!

Martina Eicher, projectmanager

You wouldn't think so, but this is really a book to take with you on holidays. The stories fit together beautifully and logically. You get into a flow and it invites you to read further. Very fascinating! Despite the serious content, it is an enjoyable read. Before you know it you'll have finished reading it in the comfort of your own home.

Maarten Hamburger, mortgage advisor

The book reads away easily and is confidence-building, because it mentions so many stories of experience. So now and then I sat reading with a smile on my face, because of recognition, but also because of the author's own story. And her sense of humour, that is important in what are considered "heavy" stories. Some lightness is really appreciated.

Karen Sjouke, body oriented therapist

I read the book in one sitting. That says something, I think. The storyline is very pleasant, with information about tapping, hypnosis and in between her personal story and the experiences of others. A nice way to read. I have to say that certain parts really touched me.

Veerle Zirkzee, customer advisor

Impressive stories and a special storyline with the author's own stories in between. I see important elements, such as the subconscious, being hypnotised by the environment, the storylines in the family, forgiveness, letting go, self-recognition and self-worth. Beautiful book!

Ina Oostrom, hypnotherapist and trainer

A clear vision on how to crack the system and the blockages and above all the good effect of all that. Jeannette makes herself vulnerable and to me personally that's the most beautiful part. An A with a stylus, I would say. (That's a Dutch expression for the best score you can get at school.)

Frank van der Zwan, casemanager social security

The directness with which the author tells her own story struck me. It made me cry really hard. I admire how openly she shares that. Which also immediately makes it clear: It is a story. You are not your story. You are the story you want to be.

Chungmei Cheng, HSP coach & rebirther

A truly fantastic book in which the author explains in a very down-to-earth, very knowledgeable and straightforward manner how past experiences, negative messages and negative self-images can really bother you. She describes how you can let go of the emotional pain and replace it with an empowering story. This allows you to enjoy life again from your strength and positive qualities. Jeannette offers you a helping hand with a technique that is as effective as it is simple: tapping. And I can only fully agree: it works.

Hans van Hilten, program director

"One who looks outside dreams,

one who looks inside awakens."

Carl Gustav Jung

For my daughter and my mother. My daughter is my strongest driver to let go. Every day again and again. What you don't release you take with you and you pass it on. To her I only want to pass on the best of the best. That's why I learnt to let go. And with her I learn how to do that and I see that I can.

My mother is my training ground. She inspires me to let go of beliefs and habits, which I don't need anymore. She has carried me and deep inside I've always known that her love is limitless.

Table of contents

Foreword

It seems surreal to be writing these words of introduction to this book on a topic that has been unbeknownst to me, the driving force of most of my very existence.

It's for that very reason, I feel honoured to firstly be alive and able to write these words of encouragement for anyone else that is navigating their way through the unravelling of trauma in any form, after finding my own way through and out of a lifetime of trauma with very little help from the professional community.

Secondly, because most people believe that trauma is something to be endured rather than understood and released as we gain the wisdom and insights it has to offer us. Once we can change the way we approach it, we are able to understand that it had something to offer us that inevitably directs us to a higher understanding of who we are and what we have to offer the world.

Finally, because this topic is so misunderstood and treated as a life sentence, rather than an opportunity to turn and see the incredible amount of strength and empowerment that each of us hold within ourselves when given the right environment to release the aspects of the experiences that our mind has deemed dangerous.

Jeannette is one of the most authentic and honest people I have ever met. We met years ago when we were both volunteering at a long-term substance abuse centre in Hawaii. From the moment we met, I knew I had met someone who was as dedicated to healing trauma as I am.

The thing that stood out to me about Jeannette from the very beginning, all those years ago, was her ability to bring fun and laughter to a situation that was otherwise considered very serious.

Like myself, she can be as blunt as a kick in the teeth, but that is part of what I love and respect about her. You never have to guess what Jeannette is thinking. I've never had to sit back and wonder about how she feels about me, someone else or a situation.

What you see is what you get and that is such rarity in a world that is full of double talk, self-serving and ambiguity. At the same time, Jeannette's love for life and the enjoyment of each moment is present in every area of her work, allowing her to bring that lightness to some very serious and heavy topics.

We've shared a lot of meals, trips to the islands, time at the beach after some very serious sessions with the people we have worked with over the years. I have worked with countless other practitioners over those years, Jeannette is what I call the real deal.

That's all to say, I have also spent a lifetime of watching people, looking for authenticity, honesty and shared values. Jeannette is someone that I trust to go to when I want an honest answer. She is the person that will tell me the truth whether I like it or not and make it humorous at the same time.

I've watched Jeannette navigate some of her own traumas through the various seasons of her own life; easy times, hard times, life changing events, including times that would knock most people completely off course and come through it all with her humour intact. Sure, as a seasoned practitioner, we automatically think that she would have, yet my experience has been that there are many practitioners out there that know the theory of what we are working with, but very few walk their talk. Very few have had their boots in the trenches, and as much as they want to help, can quite go all the way with their clients.

That is what I would want for anyone entering this journey of self-discovery to have, a guide that has already travelled the trek and knows the twists and turns that show up along the way. That has first-hand knowledge on how to not only navigate the journey, but how to successfully get to the other side and have it been an experience that uplifts you.

I take Jeannette seriously, because I have seen and experienced her work firsthand. I have seen clients that have suffered with a lifetime of trauma walk away from a session with a lightness that they have not ever experienced before. Jeannette's ability to understand how the mind processes trauma and codes those events is crucial to creating the change that you are looking for. She questions nearly everything. She thinks and processes from principles first.

For Jeannette, this is not just a job, this is her passion. She has a light approach to heavy subjects and makes getting rid of stuff very easy. Her number one talent is taking difficult things and breaking them down to a more simple and easy understanding way. She has taken a lifetime of work and broken it down into simple concepts that you are able to apply firsthand and create change in your life.

She thinks holistically and long term. Most important, she doesn't take herself too seriously.

This book will give you a good taste of what living from a place of freedom really looks and feels like. So, pay attention, what you are holding here is a way out of the pain and frustration that has kept you stuck up until this point.

Follow her advice... but only if it holds up after scrutiny and testing in your own life.

Consider everything... try it out, don't take it as gospel, check it out and see how it holds up in your own experience. Jeannette would want you to challenge her, if you bring your A-game and actually put to practise the principles she is sharing here.

Jeannette has changed countless lives for the better and if you approach the following pages with an open mind and a level of high curiosity, put the pen to paper, she might just empower you to change yours.

Kim Jewell
Behavioural Specialist
Trauma & Anxiety Expert
Queensland, Australië

Bonus material

In this book you'll find bonus material.

- ★ It helps you deepen your knowledge about this subject.

- ★ It gives you more insight of what it can mean to you.

- ★ It gives you tips to put into practice.

All the bonus material is free. Pay attention to it. There's a reason I made it. Some instructions work better with a video or a picture.

Everybody who holds this book in hand can download the bonus material from the website[1]. On some pages you can only get in by giving your name and email address, but you only have to do that once.

---

1) tapandletgo.com/

Introduction

Trauma is a word that may sound heavy to you, but it is just a Greek word (τραύμα) and it means 'wound'. And just as every child scrapes a knee when it learns to walk, what the Greeks call physical trauma, every human being acquires emotional trauma. In the first years of their life. And that is fixable. The sooner the better.

Unfortunately we're not taught this. On the contrary, we try to hide it and to forget it as soon as possible. But when you don't clean a small wound it starts to fester. The same goes for what we call psycho-trauma. Psyche (ψυχή) is the Greek word for inner. The almond paste inside the Christmas bread, the Greeks also call psyche.

Solving psycho-trauma is also a matter of cleaning up. Because before you know it you're given a diagnosis and drugs. Or you start using drugs or booze by yourself to not feel what you feel any longer. But this detour will cost you years of your life and a lot of money. There is a faster way. You can get rid of it!

Who is this book for?

This book is not for those who want to get the solution from outside. It is for everyone who has already tried that and didn't find it there. You may ask for help, but the real answer to your questions is inside you. And when you want to find it, you will. It's not even that difficult and when you love yourself just a little, it is also good fun.

There's a little courage needed for this, so if you don't have that, I suggest you close this book right now. A little bit of courage to once more face what you still remember of your miserable story (trauma). It takes a very short time and the result is worth the effort. A burden will drop off you literally and figuratively. Instead of letting this simmer beneath the surface for another eternity, you learn to relax and release it. Get rid of your rubbish, roots and all.

Why would you listen to me?

I discovered at an early age that being antisocial has great benefits. I was not such a friendly child. I was very sweet, but I didn't believe that all the adults around me deserved my handshake or for me to be kind to them. Sometimes that was awkward, especially for the adults, but this way I could stay true to my inner voice. Deep down inside me I didn't care what people thought of me and so I kept myself safe. I wasn't able to keep this up all my life. But the more I recover that power the easier my life becomes.

For more than 40 years I've been helping people to release learnt misconceptions about themselves. And from doing this work a method has evolved that works well. It contains a selection of favourite techniques I've learnt to apply which, in combination with my style and skills, appears to be effective. In recent years, I've helped people from all over the world with this method. I teach them to relax and release trauma from the past and to listen to their own voice. This year I will be 63 and because I will not be around forever, I aim to share in this book just how I do it so you can learn it too.

Over the past three years I have recorded my sessions, to get an idea of what I do exactly and how I obtain such good results. People experienced great relief and often asked me: How do you know that? What do you do exactly? And often, after a session, I was unable to answer that. But even so, I knew that I was able to and had to explain it. Every session is different, but the procedure is basically the same. I don't want people to depend on me. I want to help more people and I want it to be easy.

What do you get in this book?

We are inclined to always want to learn more, especially when we believe that we're not good enough or that we don't know enough. Go on then! Do one more course, another training, another therapy. Nothing wrong with that. I did it for years and learnt a lot. But in the end, what I find most interesting is what I find in myself. And it's also rather unique. When you really start expressing what's inside you, you become truly original. So the first thing we have to do, again and again, is relax and release it all. Stop overloading yourself. Notice what's already there and get rid of what's not useful. This book is about HOW you can do that.

Words like 'trauma' and 'hypnosis' have been scaring the hell out of people for years. I guess that's not surprising, but it's a shame. To end this misplaced fear and confusion, I've asked the help of 4 specialists. Each of them will explain a part of the terminology in their own words. With all four I've had a good conversation and processed their words and advice in this book. I will introduce them to you briefly in chapter 1.

I will describe the first example from my practice in chapter 2. Maria's story and her fear of flying and how its cause appears to have nothing to do with flying at all. Just one session was needed to understand this and to relax and release it all. And so she was able to fly to her heart's desire. She wanted to be at home everywhere. In chapter 2 I will invite you to take the first step to find out what it is you need to let go of.

In the 3rd chapter I share the story of Jules, who was heavily depressed for years. As an actor and director, he was awarded some famous prizes. But he was still living in the world of the child that was not seen. This prevented him from realising how he moved people and how much he was loved. You will also read about how we stay afloat by drowning out our pain with judgments.

Religion is discussed in chapter 4 in Sigrid's story. Religious or not, we all believe in something. Besides God or our country, we either believe in ourselves or we don't. We are told exactly who and what we are. And when we grow up, we start questioning everything. This chapter is also about acceptance of ourselves. About taking responsibility, about love for ourselves and for who and what we are.

Chapter 5 tells the story of Jente. Albeit with the best intentions, Jente was forced to eat when she was admitted to a specialist clinic even though initially, her anorexia had nothing to do with eating. I'll write about diagnoses that don't achieve anything and allow people to suffer for years. Once more, I'll be asking my less popular questions about our history. Asking questions over and over again takes you further, rather than blindly accepting all those platitudes. Relax and release all the stories you were told.

In chapter 6 you read how Abel moved from darkness to light in just a few sessions. He started to recognise his own worth, got moving, built his nest and fulfilled his deepest wish. And you also read how I was given a helping hand to release an awful lot of stuff in one go. And how liberating it is to give in to pain and effort before letting it all go.

Antonio's quest is explained in chapter 7. He wanted to escape, but didn't dare to leave the house. Like me and many others, he was a trauma survivor. A successful violinist. But after years of touring the world, he found himself totally stuck. He was back living with his parents just trying to survive. Acknowledging his emotions as well as a different perspective enabled him to go out in the world again.

In chapter 8, I tell you how you can make a radical decision in a single moment. I'll tell you how my sister realised what she had to release, the moment she held her first grandson in her arms. She took off her 'survival suit' of 30 kilos of excess weight and once again became who she really was. 'Grandma fun' says her little grandson. About how we create new experiences by releasing old stories, anger, fear, sadness and pain.

Finally, in chapter 9, I describe my hardest episode of letting go. The story of my stepfather. Once, I wished for him to just slowly rot away. Miraculously enough, that really happened. But I did not see any of it. Fortunately. And it doesn't make any difference to my life. Safety is found within yourself. That's what I've learnt, thanks to him as well.

And that has helped me a lot. Count your blessings. It's good for your heart.

Chapter 10 is dedicated to the procedure. How do you do it? Step-by-step you gather what you need in order to do the TAP and LET it GO. Don't miss out the bonuses with links to videos and additional information on my website.

Tips on how you use this book

You can just read it, of course, but at the end of every chapter there are some questions for the reader. It means that while reading the book, you can prepare for the process that I explain in the final chapter. This is one of my ways to tap and let go. It is not sacred and it is constantly evolving. Take advantage of it!

- ➤ Buy yourself a lovely notebook and pen. I can assure you that writing by hand has a better outcome than typing on a screen.
- ➤ No need for technology when you embark on this type of self-examination.
- ➤ Sit down in a comfortable place, where you won't be disturbed.
- ➤ Take 5 minutes to answer each question. Set a timer and do nothing else in those few minutes.
- ➤ Write as fast as you can and do not pay attention to punctuation or spelling. As long as you understand what it says, it's good enough.

When you want real benefit from this book, like changing your life, do these exercises and be truly curious about yourself. Don't think you already know everything, because much of what you think and believe about yourself is learned programming. It is based on the values and beliefs of others.

Take time for it and take time for yourself. You can read all the books in the world, but when you don't put it into practice there will be no results. We all have to do the work ourselves, because nobody else can do it for you.

You are unique and special and worthwhile. You deserve to live rather than survive.

Check out the bonus material. It's worth it. Know that you are worthy enough to do it.

Now that you know what to expect, I'd say: Go for it!

Have fun!

Jeannette van Uffelen
The Hague, The Netherlands
April 2024

1. If you hold on, you're going to die

Shortly after I published my first book 'Undercurrent' I gave Paul de Blot a copy. And three days later he sent me his review. At the time, he was a professor at Nyenrode Business University, where a magazine was published every quarter. It was a very thorough and enthusiastic review. He thought I wrote well and had a clear style, so he asked me if I wanted to write for the magazine. It was about Business Spirituality and I did in fact contribute articles for a while.

Paul said to me, "If you're going to write another book, let me know. Because I'd be happy to write a foreword in your book. I am a professor and doctor and people find that interesting. So that might help you." He gave me a big grin and looked at me with wide open eyes, mischievous, enthusiastic and wise. It took me a long time before the second book came out. And Paul passed away in December 2019. He was 95, so I can't blame him. I don't ever blame anyone for dying and I wish everyone a peaceful ending.

Instead, Kim Jewell wrote the foreword, but Paul de Blot deserves a place in this book, because he often discussed the theme of 'letting go'. He said:

"If you want to hold on, you die. If you let go, you stay alive."

He had learnt that in the Japanese concentration camp, which he survived for five years. For the last year he was in solitary confinement. "And there", he said, "I had almost given up, but when I heard someone knocking on a wall somewhere, I knew I wasn't alone. So at that moment I said YES to life again."

You can't really describe the theme of Letting Go any more powerfully than that. Without a doubt, Paul de Blot was a person of clear statements. Straight to the point. Boom! Bang! I love that. Why would you make it harder? Just say it, right now. Not to hurt anyone. On the contrary. Always with kindness and compassion and an open heart. Because there is nothing as liberating as a clear and frank conversation.

My first book was commissioned by the publisher and I wrote it in 3 months because I had been given a deadline. In fact, it took me just a few weeks to write and in the remaining time, I exhausted myself with its finer details. I enjoyed the writing process immensely and it flowed easily. The book was well received. It was considered very readable, because of its 'light-hearted style' about a tricky subject, like Human Resources Management. Why haven't I written at least one book a year since then? If I knew I would have done it. Too busy, too distracted, too insecure, too ... you name it. But now, I write because I've let go of what was holding me back. And that's what this book is about.

Taking yourself and others seriously

Kim Jewell says in the foreword that I don't take myself too seriously. I think that's a great characteristic. If I give too much importance to my own opinion and myself, the fun is lost. Then life gets hard and before you know it, you get into a fight. I can have strong opinions, but I'm also able to give an immediate rebuttal to every sentence I utter, often before I even say it. That can be exhausting, but it also makes life simpler. It means that I am not desperate to 'be right'. At the end of the day, I think we all just do what feels right. No one has the ultimate answer and we don't necessarily have to agree with each other.

If you don't take yourself too seriously, you make being and working together a lot more fun, more interesting, and a little easier for everyone. And so 1 + 1 became 5, rather than 3! Having conversations with 4 specialists has given this book added value, which has done me and my readers a service. I have also experienced once again, how much fun it is to exchange ideas with professional brothers and sisters about this fascinating work. And it strengthened my conviction that this book had to be finished as soon as possible. "Good timing!" wrote one of the co-readers.

All four specialists are, like me, certified hypnotherapists and some of them are also stage hypnotists, masters in Emotional Focussed Transformation (EFT) or Neuro Linguistic Programming (NLP). They have been providing training in this field for many years. Two men and two women, two Dutch, an Australian and a Belgian. I know their work and admire their drive to constantly bring to the attention the power and usefulness of hypnosis and related techniques. Groundbreaking work that they do with great passion. There's a lot of similarity in what they say, so I don't quote them all on every question and they each have their own specialities.

Ina Oostrom I met almost 20 years ago. Authentic and passionate, with many areas of interest, she immediately impressed me. Ina has run her training institute for many years, providing various courses. She writes about hypnosis and her ongoing research including the use of hypnosis for pain relief in operations. She has also created a community website, where people can get acquainted with hypnosis for free.

I met **Kim Jewell** when we both volunteered at an addiction treatment centre in Hawaii. We gave sessions together and that way you quickly get to know each other. Outside of work, we also like to hang out with each other. Because Kim knows me well, both in my spare time and how I do my work. I also asked her to write the foreword for this book. Kim is a hypnotherapist, NLP and FasterEFT master. She has developed her own online 'Systems of Change' training, based on the work of Stephen Heller and others.

Martijn Groenendal made it easy for me. After working a lot abroad, he started his training school in The Hague, my town. I have taken part in his training courses there and we also worked together when he brought foreign trainers to the Netherlands. Martijn started with NLP at the age of eighteen and has been delivering NLP training and variants of hypnosis and hypnotherapy for many years. He organises his international bootcamps at his own academy and as part of Igor Ledochowsky's team at the Hypnosis Training Academy.

And I also met **Rob de Groof** in The Hague when, after all my training abroad, I was looking for something closer to home. Rob is the founder of the largest hypnosis institute in Belgium and offers a wide range of long and short training courses in this field. Rob is a real entertainer and has toured with his own stage hypnosis shows for years. If you're lucky (and I was) he does a show during the training breaks which he calls the 'Trick of the day'. He no longer does live shows, because, in addition to managing the centre in Belgium, he is fully engaged in his growing international hypnosis school.

The details of these specialists are in the back of the book. And you can listen to the four conversations in my podcast.

Tap & let go, why the title?

What do I actually know about tapping and letting go and why is that a good thing? Good questions. I know a lot about letting go and I am a master at it. Until I started the podcast, I had never before focused so consciously on it. I launched a podcast and bought a domain name with this title (in Dutch). I made a corporate identity with a photo and logo. And when I saw it around me every day, I realised even more that this has been my theme for years. When you read 'Let it go' you immediately have an idea of what it is about.

I've been practising tapping for ten years. It is a very simple variant of acupressure, whereby meridian points are lightly tapped with the fingertips. FasterEFT (Emotional Focused Transformation), which is also called Eutaptics and Clinical EFT (Emotional Freedom Technique), ERT (Emotion Replacement Therapy) and Scientific EEG (Electro Encephalo Gram) Hypnosis where brain waves are measured during the hypnotic sessions. At the end of the nineties I

19

was trained as an ICT (Information & Communication Technology) specialist. Before my technical training, I was already familiar with NLP. Language is a powerful tool for (re)programming. This also applies to the human brain. And if you do it smartly, it can make a big difference. It's fascinating and fun!

Every day, I immerse myself in behavioural, cognitive and physical methods, hypnosis and hypnotherapy. My favourite phrase is, "Whatever it is, it's time to let it go." I have uttered this phrase many thousands of times. Alone, in groups and in many sessions. This phrase has become dear to me. Every day I understand and experience more just how powerful and helpful these words are. With the emphasis on "Whatever it is". You can let go of everything, whatever it is. I help people to let go and practise it myself every day. Because letting go creates relief. You literally feel lighter, because you have less to carry around. It has become my daily activity, my life actually. So obviously, this is also the theme of my podcast. The name came to me spontaneously and naturally.

Most of my work is in English but I decided to write this book in my mother tongue. Close to home and close to myself. And right after that I translated it into English. Even though I can express myself well in English, Dutch is still closer to my heart. It is the first language I spoke as a child when I learnt everything from my parents and grandparents. It is the language I was raised and programmed with. So I chose my first language and, right away, the title popped into my head. In Dutch it is 'Lekker los laten'. 'Los laten' means let go or release. 'Lekker' is an almost untranslatable Dutch word, used in many contexts. It describes something tasty as well as comfortable feelings, nice weather, a cosy home, a comfortable couch, a fitting chair, an attractive body, et cetera. It sounds good and it fits perfectly with what I want to achieve with this book: I want to make it easier. Simple, pleasant and therefore 'lekker' easy to read and 'lekker' to do.

The conversations in my podcast are about 'Letting go' and it's also what I'm going to share with you in this book. Is there an easier way? It seems like everybody experiences stress, but is that really necessary? I don't think so, actually. So, what can we do about it? Not so much doing something about the stress, I guess but instead, taking care not to experience stress and not to hold on to it. But life is unpredictable, so you can be shaken up by something at any moment.

That's why I want to teach you a way to release stress quickly. Something you can immediately do yourself at any time. And I want to teach it to you in a simple way. So no jargon. And no pages full of to-do lists, planning plans and more stress-increasing pressure. No need for forcing yourself to sit quietly in meditation when you haven't got time for it. But what then? Read on.

The stories are here already

The past is gone. It is over! You cannot turn back the clock. And often, we wish it had been different. But that's just not possible. It can't be done! We cannot change the past. And you don't have to, in order to carry on with your life. Instead, you only need to change your perception. You are able to do that and I'm going to teach you how.

This book is filled with stories from the past. I am unable to explain the process without giving you examples. These are examples from my history. For a large part, I'll describe what used to be my old perception. How I saw it for many years. The way I used to experience my own stories and how I've believed and talked about them for years. During the writing or when reading aloud, I sometimes burst out laughing. This surprised me, because I've known those stories all my life. But still, I had to stop the recording, because I had to laugh so much at my family's sayings and expressions. As a child, they were impossible to understand. But once the emotions were removed from the stories, I saw just how bizarre it all was. And when I sat down to write and then re-read what I had written, I often thought: how did you come up with that? And that often makes the writing so easy for me. I don't have to make it up. The stories are already there. All I have to do is to write them down.

I started a blog once about my family stories, because my parents and grandparents had a great way with words. Often coarse, but really funny. It was part of our survival. Mine too. And I've often heard this from friends, colleagues and clients. At every assessment, evaluation and farewell party it was mentioned: my weird sense of humour, spot on with great timing. I don't know any better. I see humour in many situations and would find myself saying something before I realised it. I needed that in the past and I still need it. I don't always express it out loud, because that saves me a lot of trouble. I don't like to explain it. It's fun. That's all. People sometimes get easily offended and that's not my intention. I find many things funny. And people who take themselves very seriously I find hilarious. I don't expect anyone to agree with me, but I cherish my craziness and my weird perception. It's what's kept me going.

To heal myself I had to let go of old stories and I wanted to forgive everyone. That was quite hard at the beginning, because my stories had shaped my identity. I've also let that go. I wasn't that nice a person but I only realised that after I let go of my beliefs.

I became a nicer person, for myself as well. And so, I have rewritten my history, which I'm still doing and will keep on doing. As a result, I've started to look over and over again at my parents and grandparents. Every time you tell a story it changes. Because every time, you change your perception a little. And more and more, I see and experience love in those stories. And I notice that I've forgotten a lot and I like that. Stories I repeated over and over in the past are no longer that interesting. And more precious memories have come floating to the surface.

Cold chicken soup

It has been ages since I last ate chicken soup. When the soup is cold, you can make a hole with your spoon in the layer of fat that's on top. I liked to do that when I was a child. It was also kind of dirty. Old stories are like that fatty layer. You can't look through it and so you're unable to see the delicious stuff that's floating below.

When you heat up chicken soup it starts to move and the fatty layer disappears, so that you can see everything again and taste it. Something very similar happens in the search for underlying emotions, during the process of releasing and rewriting. You rediscover nutritious stuff, that you thought was lost, blinded as you were by your entrenched beliefs. In a way, you were very hungry, because you couldn't see the food.

This book also contains stories from other people's past. Stories from sessions with my clients. All names are fictitious. And I changed some details to protect their privacy. Some of these stories are on my website, because some people wanted to share how quickly and completely you can change. They experienced how, after going through the process, you are able to reach your full potential. In the stories they initially sound like victims or even the culprit of their story and that's why the original story bothered them. I also tend to leave out a lot of unnecessary detail.

This book is going to be read by empathic types like you and me. And they are going to feel the stories. If that happens, I suggest you tap along while reading. I explain in this book how you do that. You'll find the instructions in chapter 10. When you follow them, you will immediately release the emotions.

The praying mantis

Three summers ago I walked with Susan along the beach of Agios Ioannis, Saint John, on the Greek peninsula of Pilion. She saw a praying mantis hitching a ride on the rim of my hat. The day before I'd had one sitting on my bed. When I stood still, the praying mantis walked all the way around my head. Susan took some photos and told me:

"The praying mantis appears when we have our life filled with too many things. As a result, we don't hear our silent inner voice anymore. The praying mantis insists that we take a step back. In other words, it is time for some simple meditation. It is the only way for us to come back to our truth. This spirit animal always visits us when we need some peace and calm in our lives."

I don't know if this is true of course, but what is true? I guess truth is a subjective concept. I speak several languages, but I've never talked to a praying mantis. I don't understand them, but that doesn't mean that they don't talk to me. I appreciated this beautiful story and felt honoured to be visited by this tiny creature in my temporary home as well as on my hat. And the advice, as Susan put it, couldn't hurt at all. How much evidence do you need to take something onboard? How it feels to me says it all. And to trust that feeling, to learn from it and to live according to it has made my life easier. That's how I give my life meaning. And that energises and motivates me every single day. It's a matter of perception and we can choose which story to hang on to and which story to let go of.

Daring to let go is empowering

A lot of stress comes from the fear of loss and change. And from fear of letting go. We often don't even dare to let go of what makes us deeply unhappy. Because we just don't remember what it is like to live without that shit. A lot of trauma originates from the first years of our life. But holding on to fear negatively impacts your life and causes stress. Often, it arises mainly in your head, as a thought. We are able to imagine the weirdest stuff in an instant and scare the hell out of ourselves. And if you're not careful, you start to believe it and you adjust your life based on the fear of what you think might happen.

Life is unpredictable and you don't know what is waiting for you. When you trust that it's always getting better, you don't have to hold on to anything. That doesn't mean to say you always have to strive for more. Or that you're not allowed to grieve when you lose someone. Nor does it mean that what you have with people or things is unimportant. On the contrary. You have to cherish what you love and keep it close so it stays with you.

But knowing that you are able to let go gives you peace, relaxation and trust in yourself. And that is what you need in times of imminent change.

How about you?

1. Is there something from the past you would like to let go of?

2. Have you ever been forced to let go? I.e. because of death or divorce?

3. Or do you let go (maybe too) fast, and later on, start doubting?

Take some time for yourself and get your pen and paper. Write the answers to the questions above for yourself and feel what it does to you.

Or don't do any of that at all and continue reading. Remember, it's all about free choice!

2. There's a reason for everything

Having a childhood trauma doesn't always mean that you had horrible parents. You can have a trauma without having experienced violence. I once had a young houseguest, who was happy when she arrived but became sad as it got closer to bedtime. After asking her a few questions, it was clear that she did not want to stay overnight. She missed her mum. So I called her mum and the girl slept in her own bed that night.

When the fear of flying is not about flying

Maria (52) wants to visit her daughter who moved to New Zealand, but she is dreading the trip. She has a fear of flying. She flew once before, but that was quite dramatic, so she decided never to fly again. But now she has a very good reason to change that. She wants to see her child and her grandchild again. One session turns out to be enough to solve the whole problem. And I will briefly describe below how a session with me works. It is never the same, but I do follow a procedure, albeit with a few variables. I adapt what I do, depending on what I sense at the time.

I ask her to describe the problem, as briefly as possible. What do you want? What do you want to get rid of? "I always have such a panicky feeling. I want to get rid of that. I want to relax wherever I am. I want to feel at home wherever I go and I want to enjoy travelling. All the time I'm thinking: Will I be able to get back? And I constantly think about what could go wrong. It's always the same."

First of all, I ask Maria in great detail what makes her happy. "When it's easy, being in nature, making something, cycling, walking. The beach, the forest, the trees, the energy of the forest. Then I feel supported, absorbed, one with nature."

Next, I ask for more details about what she wants to get rid of. "I always get so restless, nervous. What if this happens, if that happens, if I get lost. Panic, then I find myself without my own house, without my stuff. It keeps coming back, I have a hard time letting it go, I dare not go and visit my child. I find it too difficult, all the worrying, I think it's so stupid. I don't want to leave at all."

At home everywhere

So I ask her what she wants instead of the above. How do you want to feel after your fear of flying has disappeared? "Then I'll be at home everywhere. I'm Maria and my home is everywhere. Wherever I am, I'm relaxed, I'm free. I want to do something with people." I let her sit comfortably and tell her to close her eyes and take a deep breath in and out. I ask her to go to the feeling of panic in her mind. Where do you feel it in your body? "In my chest, I can't breathe properly, I'm curled up, there's a tightness in my arms." I tell her to make it stronger. Feel it, see it and know it's true.

Have you felt, seen, experienced this before? "Yes." I ask more questions and take her further back in time to previous experiences. And then, to her own surprise, she remembers a sleepover at her aunt and uncle's. She was very much looking forward to it, they were lovely people, living in beautiful surroundings. But suddenly, she really missed her own home, her parents, her familiar surroundings.

During a session we use tapping on meridians. And I help her to let go of all unwanted emotions. And at the same time, I repeat all the desired experiences. In addition, I repeat what she has said makes her happy. While I talk, I tap on her face or stroke her head, arms, hands.

And that's how you change painful memories. You let go of the associated emotions. And you immediately replace them with what feels good, which gives you a relaxed feeling. You can't be afraid and happy at the same time. I will tell you later in this book how exactly you do that. Timing is crucial. The right information at the right time.

Then I help Maria imagine that she is going to fly. That she sits in a plane for hours, over mountains, countries and oceans. We do the whole preparation. She tells me what she does and I help her to feel it, to see it, to hear it, to experience it, to taste it and even to smell it. And as soon as something comes up that causes panic, we tap on her face. I also teach her to do this herself, so she can use it at any time. If necessary, on the trip, on the plane, on the road. Wherever and whenever it's needed. Finally, she's able to imagine sitting relaxed on a plane and even falling asleep. We add everything that can strengthen her confidence in herself. She and I, using her own words as much as possible. Because these really resonate with her.

The body does not know the difference between an event in 'the here and now' and imagination. So, with a little imagination, you can give yourself an experience. It's easy and anyone can do it. We do it all

31

day long. Unfortunately, we often (subconsciously) do not choose the best fantasy. Instead, we worry and create stress in our thoughts and in our bodies. This way, we reinforce the fear and disbelief in our possibilities.

She couldn't find it anymore

You can change. It's not that hard. But you have to make the effort in order to learn how to do it. Because so much of what we do happens subconsciously. We have been taught for many years. We are trained, programmed and hypnotised. We have a lot to unlearn.

A few weeks later, Maria boarded the plane and flew to the other side of the world. Of course, it was exciting, but she also experienced her adventurous side. And it was wonderful to travel there and go into nature. At first, Maria found it hard to believe that her fear of flying had been resolved. We never learnt that we can learn to let go so easily. That indeterminate, underlying feeling of panic had disappeared. She couldn't find it anymore. At the end of the day, it had nothing to do with flying at all. It was her fear of not being at home. And being at home has everything to do with feeling good in yourself. That was her real wish. Feeling at home everywhere.

We know everything

Do you believe you were born insecure? I can hardly imagine it. I have seen, held and cared for many babies. I once worked with young mothers and their children in crisis care. I am convinced that babies' insecurities are related to the circumstances of their birth, including the months spent in their mother's tummy. We already have an awareness in utero. We sense whether or not we are wanted. We know if mum is comfortable in her own skin and if she is looking forward to our arrival. And whether she trusts herself, the world and life or whether this is not the case. And then you are born and that's not always easy.

King of the world

The majority of the babies I've watched growing up were not insecure in themselves. Certainly not in the first years of their lives. On the contrary, they considered themselves the centre of the universe. At least, that's how they behaved, regardless of whether or not they knew what the universe was. If they want to eat, they let you know. Or if there's something else they don't like, they'll express it and you have to guess what's going on. If a baby feels safe, is seen and heard, there is no trace of insecurity. For the past two and a half years I've had such a beautiful being in my life. He's in my care at least 1 day a week. It's an amazing experience again. And now that he's talking more and more, it's even clearer how the world revolves around him. And how sure he is of himself and trusts that I am there for him.

He enjoys coming here and sometimes asks his mother: "Go to Nettie?" Then I usually hear him chatting outside and calling my name. He walks in with an I'm king-of-the-world look. He looks straight at me, his face all joyful. And then he walks past me to turn the whole house upside down again.

We have a busy schedule. First of all, we eat, because I love to watch him enjoying my food. And because he doesn't have time for that at home. "No food. Go Nettie." "Kiwi tasty, pancake tasty" or "no pancake". Drawing, cycling, seeing the cows. Farm, watch chickens or eat crisps, prepare a bottle of milk, beach, piano and so on. I have to make choices, depending on the weather and how well he has slept, but he just wants everything. He wants to do everything and is convinced that he can do anything. When that turns out not to be the case, he sometimes falls. And then he gets angry and doesn't want to be comforted or he screams and throws stuff. Or he cries and then I may comfort him sometimes. The frustration lasts for a very short time. Because usually, he tries again about ten times or he lets it go and proceeds with the order (or chaos) of the day.

No clock and no time

Small children do not have watches and clocks, so they live in the here and now. They eat when they are hungry or when they see something that appears very tasty. They play as long as it's fun. And when something else comes along that seems more fun, they switch over quickly and abandon what they were doing. They dive 100% into the next thing. Don't tell a two-year-old, "First you have a nap and then we go to the zoo." Because they remember the last part of the sentence, they are immediately ready for the zoo. They have no sense of time, before or after. And they don't sleep a wink because of pure excitement, if they know how much fun a zoo is. Very recently, I asked him to choose an ice cream at the local playground. I took two ice creams from the freezer and asked, "This one or this one?" He chose the second. I put the first one back and took a third one and asked again, "This one or this one?" He chose the second. I put the first one back and took a third one and asked again, "This one or this one?" He chose the second one again. I did that again and I could have gone on like this for another hour, because children of that age choose the last option. Every time.

That's why you shouldn't ask a two year old what they would like to eat tonight. You have to make that decision. No need to burden them with that question. And yes, if they don't want the same thing as you, they can make a lot of fuss. Then you have to explain it all to them, preferably at the level of a two-year-old. You don't give them an encyclopaedia or a 10-minute YouTube video with explanations. If a child doesn't understand it, they can get angry out of sheer frustration and then throw something or start yelling. That's not fun, but just imagine yourself in his or her position. Grownups don't understand you. You can't reach anything. You're thirsty. You're tired. But you don't know that yourself. One of many daily frustrations.

But in general, small children are easily satisfied if you take your time with them. When they know they are wanted and loved, regardless of what they do. When they enjoy themselves doing something or making something they can lose themselves for hours. This is well before the start of feeling criticised and compared to others. Small children come to show you their drawings, paintings or other work. And if you're unable to see it yourself, they'll tell you it's a fish or it's daddy. Or that it's you, with the big head and the short legs, without a nose and with three hairs on your head and a pair of glasses on your stomach.

There's a reason for everything

As a child, I spent hours drawing and I lay in the water for ages, especially under water. I really enjoyed learning to swim. As soon as I was able to swim, I was always at the pool. I jumped off the high and low diving board. Many times in a row. I did handstands under the water and swam many laps, staying below the surface for as long as possible. There were summers when I had an ear infection, because I was underwater too often. Even more than in the pool, I loved being in the sea and that is still the case, all year round even.

There is a reason for everything and that includes uncertainty about yourself and your performance. But there are also achievements, tasks, activities, you do with the greatest ease. Things you do effortlessly, with ease and without shame, as long as no one sees, hears or knows about it. So you sing loudly in the shower with a clear voice and the whole neighbourhood can hear it. But during the choir rehearsal you notice that everyone around you sounds so much better than you.

There was a time when you made something with ease and considered it to be okay. Like I spent days drawing or building something out of bricks at the construction site down the road. I built us a house with a window and a roof. And of course, we built sandcastles, knowing that the sea would wash them away that same evening. That was a shame, but never a reason to not do it again the next day. And I built crocodiles, turtles and so many other animals out of sand. And I built boats and cars around my little brother. And a generation later I built boats and cars and dolphins around my child and around other people's children and I will do that again. For that darling little boy.

What is trauma?

Martijn Groenendal:

A trauma is a reminder of a situation or period in your life, when you were unable to deal with certain people or situations. As a result, a very strong impression remained in your own brain. Something you could not deal with at the time. This can be emotional, physical or spiritual pain.

Kim Jewell:

Any time you perceive that your life could be in danger. So I often explain to people that a child can be traumatised by being left crying in a cot or a crib, if they think the parent isn't going to come back. A child can be traumatised by falling off a bike. You can have two kids riding a bike. One can fall off and think it's no big deal and the other one being super sensitive and that can be very life-threatening to that child. So trauma is any time that we perceive it could be life-threatening. And I know the American Medical Association had to change the definition of trauma back in 1995 for that reason alone because it's about the perception of the life-threatening event.

Rob de Groof:

Something has happened in the past and that is recorded on the hard drive, which causes issues. These can be very simple, but also pretty huge.

Ina Oostrom:

If seven people see an accident or seven people are in a war situation, then not all seven people will develop PTSD, post traumatic stress syndrome. It may only be one or two. And those are the people who have previously experienced trauma, as a result of which they are less able to deal with the adverse event later on.

How about you?

When you want to learn to let go, you have to evaluate your view of the past. So you get to know what is actually yours. And what is just programmed and dropped onto you. Take pen and paper and give answers to the following questions.

4. Do you remember activities which (as a child) you could totally surrender to? And do you still do those things?

5. How often do you take the time to do what you really like?

6. Do you ever do something just for yourself, which is not useful for anyone else at all?

Example: As a child I loved to draw and was good at it, just like my father. But according to the 'normal' where I grew up, it was absolutely impossible to earn money that way. That was not an option at all. Now I paint and sell icons, traditional ones as well as in my own style. Some are hanging on walls abroad. I still don't quite believe it, although I do have the proof in my hands.[1]

---

1) https://iconicpresent.com

3. Survivors are tough

Jules (49) was convinced that he could not be helped. He'd had a lot of therapy which, in the end, had not helped him at all. For two years he underwent almost daily psychoanalysis sessions for his depression. People who have experienced a lot of therapy often make it hard for themselves. And for me sometimes as well. They are totally convinced about their own miserable journey. They may have spent many years analysing and researching why they are in such a bad place. I did that too, but it didn't make me better.

Most of the time it's because of their mother. Just because she is most present in the first years of life. And if she was not present, she must have been absent, so that it's also her fault. And unfortunately, we may well have a useless father too. Depressed people always are rather unfortunate, because that's how they see life and that means it is true. The father was often absent and that's why he did wrong. Or he was very present and did everything wrong.

"There is an expiry date on blaming your parents"

J.K. Rowling

As long as you keep blaming others, you don't really take responsibility for your life and you'll run the real risk that you will not get past the complaining. And what you believe is true, so you will see proof of that all day long. So if you did not feel seen by your mother, you still won't be seen. That's how life works. You have drawn the conclusion that you are not appreciated. And without realising it, you do the same, because that has become your 'normal'. Understandable, but it's of no use to you. That was also Jules' problem.

Heart and soul

Jules is a novelist and playwright, actor and director. He has written and directed wonderful plays. And his work has been very highly regarded. It is a profession where you are vulnerable. You throw your heart and soul into a role and work on it for months. After which any sucker can obliterate you, for any reason whatsoever.

Jules has at times been insulted by critics of course but that's part of the job. He has also been recognised, appreciated and loved by critics, colleagues and his audience. For his acting, his plays and novels as well as his work as a director.

Because I also apply tapping during my individual sessions, I am seated quite close to the client. I tap my fingertips on meridian points on the client's face. This relaxes and helps with letting go. I usually sit at arm's length so I can reach my client's face. In the summer of 2020, quite a lot was going on and no one knew exactly what was happening. Jules listened faithfully to the news. He didn't know what to believe and whether or not he should keep his distance. That's why I sat at an appropriate distance from him on the bench. When I work with people online, I teach them how to tap on meridian points themselves. And then in between I give them instructions on how and when they should do this. The process interrupts the trance from which people tell their story. So that's how I explained it to Jules.

Jules had often told his stories to psychiatrists, psychologists and therapists. He had analysed himself with professional help for several years. He was used to listening to himself and going into detail. But every time you recount a memory, you experience it again. And as such, you feel the accompanying emotions all over again. Not then, but in the now, right now. That's what you elicit and you experience everything again, feeling all the pain and sadness again. So he retold and relived his memories in detail. And when I told him: "Just tap now" he would do it very slowly and it had no effect. He actually found it quite an interference when I told him to break his trance. He felt very comfortable in his story.

Itchy hands

Similarly during the second session, I was on the edge of my seat and had to watch him not doing what I knew would work. So I decided to make a face mask. At home, I could wear this face covering if I could help someone. And that's what I wanted to do. I have done crazier things than this to force a breakthrough. So I bought a pair of white soft gloves for people with eczema and itchy hands. As for my hands, I was not bothered by anything. My itching was more figurative. I cut a piece of cotton from a white pillowcase and sewed it into the shape of a face mask. I sat down in front of the mirror and painted my own nose and mouth on the mask. It was a caricature and worked out pretty well. I made a second one with wider lips and a bigger grin. I'd be able to use this to surprise him when needed.

For the third session, I explained to Jules what my plan was. And I asked if it was okay if I sat closer, wearing a mask and gloves. I asked permission to tap on his face and he was fine with that. In the middle of his 'trance', I tapped his face. And I gave him instructions. He became totally confused and that was exactly the intention. First of all, he was irritated, overwhelmed and disturbed by my interruption. He's a proud actor, who doesn't want anyone to just interrupt his performance. Next, I asked him a totally irrelevant, ridiculous question and he laughed out loud.

Jules was raised by parents who were very preoccupied with themselves. Survivors, in other words, who had very little time or space for this child. At least, that's how he experienced it and that's how he kept telling the story. To psychiatrists and to himself. He felt unseen, unrecognised, unsupported and never reassured. He had become a frightened child with an enormous urge to assert himself. He had to and would be heard and seen!

The urge to make himself heard and seen became his greatest asset. He expressed himself as an actor, author and director. He also encouraged others to express themselves. That was wonderful and much appreciated by his audience. Despite that, he judged himself, based on his experience of his younger years. He didn't matter. He wasn't important, he was never reassured and he had to figure it out for himself. He didn't feel loved. He was still looking for confirmation outside of himself, but found it nowhere.

Even love you have to find within yourself as a start. Anyway, if you're unable to love yourself, you have nothing to offer and nothing to ask for. If there is no solid base, everything leaks through and the one who loves you becomes completely exhausted and empty. No healthy person wants that. As such, you will attract the unhealthy types, who are going to manipulate you, because that's what you do to yourself.

Dance classes in the insurance package

I don't remember word for word what I said or what I did, but I managed to do what I intended. I broke through his pattern and that was a totally new experience for Jules. He had no idea, because it happened largely in the subconscious. And in the subconscious, everything goes faster. There, you let go very quickly and easily. And there the emotion is replaced by different feelings and rapid change is possible. After several interventions, which I do intuitively, I noticed him yawning and relaxing more and more. I can't usually remember exactly what I said myself. And that doesn't matter. His face softened and became younger, happier, friendlier, kinder.

The person who referred him to me called me and said, "I don't know what you did to him, but it's worked." Jules had called him and he was also unable to say what had happened exactly. He just knew that he had not felt so good in many years. He had laughed a lot and that has a powerful effect. He came to see me once more and during that session, we cleaned up even more.

It was hard for Jules to believe that he could really change for good. Let alone acknowledge that he had already changed. That his depressions were not forever but that they would disappear if he stopped creating them. Being depressed had become part of his identity. It sounds crazy, but a diagnosis always has advantages. It gives you an excuse not to get on with your life. A reason to no longer take responsibility. Noone does that on purpose and usually not consciously. But unfortunately, it is kind of encouraged and rewarded in our Western society.

There are pills you can take, which are reimbursed by your insurance. And cognitive therapies, during which you can tell your story three hundred times, without ever changing your perception. Dance lessons should be reimbursed by the insurance. And the gym and all other alternatives that make us happy and make us feel better. I would like that as part of my basic insurance package. But unfortunately, that is not yet the case. Jules told me he would talk to his doctor about adjusting the antidepressants. I don't stop anyone from doing that, because that's not my job.

I do point out that those pills are addictive, but everyone has to make that choice themselves and also choose the right time. The friend who referred him can see that he is doing much better. Jules decluttered his house and got rid of a lot of stuff. That is a good sign and, of course, a clear proof of letting go. Something he had been planning for a long time, but never got around to. Letting go of ideas, beliefs and things always creates space. If you dare to let go of what you hold on to what you hold on to can also let you go.

Eighty bags released to relax

Michael (42) had been a widower for four years when he came to see me. After a few sessions, he told me that he had finally started to tidy up his house. His mother always wanted to help him with it, but that usually ended in a fight. And according to Michael, this always resulted in an even bigger mess. I gave him a few tips and advised him to start doing it alone. He had long been convinced that he would be unable to do anything. That's how he was programmed and that had become his 'normal', so he didn't even try anymore. I told him he couldn't keep hanging on mum's skirts any longer. He appreciated the advice, but behaving accordingly was still a challenge. "You can't change your mother, so what do you do?"

My most successful clients appreciate my direct approach and do something with it. In the months before, I had lovingly helped him let go of sadness, anger, pain, frustration, guilt and a lot of old hurt. So that he could change his perception and create a more positive self-image.

But then comes the moment for the kick up the butt. Not literally though. A good listener only needs half a word. "Of course you can tidy up. You just create the urge to do it," I said. It turned out he was very good at clearing out and he had fun doing it. Furniture went to the recycle store and to the dump. He sold a few things and he filled 80, yes, eighty! rubbish bags. What do you think something like that does to a person? Of course, it perked him up tremendously. You could see it and hear it in his voice. His whole demeanour changed.

His mother wanted to help him of course, because she really loved her child. But that child was now 42. And true love can hurt. Cut that umbilical cord, otherwise it becomes a stranglehold. And if mom can't or won't cut it, Michel has to take responsibility for his own life . He has to unlearn asking his mum and then get angry when she takes over again.

YouTube is made by technical guys, who have to get away from their mother, right? And that's why they upload all those instructional videos. And hints & tips for anyone who wants to live independently. Sustainable, hipster, vegan cheese and barbecue. Everything is there! Wearing your minimalistic black turtleneck every day, just like Jobs? Everything is possible and an awful lot is allowed. You have to take the first step. At least, if you want to change something in your life, you shouldn't sit around waiting for your mother to change. Why would she?

Knowing alone doesn't help you

Even with everything I know, I still procrastinate. I know that going inside of me is really the best solution for letting go of old shit. And yet, I still postpone.

The reason for this, I also know, is that deep down I still believe what I have been taught. And it's been taught to me by people who were peddling the biggest nonsense about themselves. And about the world and therefore also about me. They did so out of ignorance. They usually didn't do it to make it hard for me. I now know that the outside world is much bigger than the world inside my childhood home. I got to know other people from other families, other cities, from other countries, continents, cultures, as well as warmer and colder countries. As my world expanded, I came to understand more. And I understood that what people said and believed in my home was not the ultimate truth, thank God.

By the way, things could be much crazier and much worse than they were at home. I was held, washed, dressed and fed from the day I was born. I was not beaten, I went to school, to gymnastics, to swimming lessons, to handball, to dance lessons, to the sea. And every day I could return home, where I shared a room with my sister and where we had our own bed. At bedtime, we bade our parents and brother and each other 'goodnight' with a kiss. There were no floods, no bombs, no hunger, to name but a few.

Open to everything and everyone

The first years of our life define us. In those years, we are still open to everything and everyone. That's when we still live in a kind of permanent dream state. We don't yet distinguish between ours and yours. We just let everything wash over us and we cope. Those first years, when we are subconsciously programmed and hypnotised, are really essential. It just happens because we all do pretty much what we know and what we've learned ourselves. We don't know any better.

All my life, I have been trying to understand my mother. Who doesn't? My daughter? Actually, she does as well. And I wish her good luck with that! The desire to know and understand my mother and father is and was a strong driving force. Perhaps this is exactly how I developed my talent. The gift of seeing through people's behaviour.

Often, when I look and listen to my mother, I don't understand a thing. She seems to be and think so differently from me. Luckily, I'm able to see her more and more as she really is and let her be. I used to want to change her. Because I was totally convinced that I knew better. I also knew how her life could be easier. But that's impossible. You can't change anyone. And you don't even have the right to. Not even with the best of intentions.

"You cannot judge someone until you've walked a mile in their shoes."

White Eagle

Well, I did try walking in my mother's shoes as a small child. But that didn't make me understand her any better. I really didn't understand how and why she managed to walk in those high heels.

She told me one day. A pub landlord whose premises she cleaned daily when she was 14 years old, explained. He made her put on shoes with heels and said, "Look, Nelly, this is what a woman's leg should look like." As he pointed to her tightly muscled calves. He and his wife were very kind to my mother and she had never received so much attention and beautiful gifts. So, until she was sixty, my mother wore high heels. Now she doesn't. She wears comfortable shoes instead. And this way, I find her much more genuine and closer to her true self. But then again, who am I? "You are different" says my mother.

50

I understood as little about her as she did about me. But she loves me a lot. I know she does. Because she often found me challenging because we didn't understand each other, she often thought I was angry with her. And I was. It was pure, heartfelt frustration. Because I always loved my mother a lot, even when sometimes I didn't want to love her anymore. Because at times, it hurt so much.

My mother talked 'at' me all my life. I was always a good listener, especially when I was little. I needed her and I didn't know any better. When I got older, I didn't have the patience for that anymore and pushed back. I would tell her over and over again how she should see things differently and how she should be different. Recently she said: "Yes, I do listen to you. And sometimes, I act on it. And that helps."

Fortunately, I am less and less inclined to interrupt and correct her. She is allowed to be who she is. She can't be anything else. All the wise lessons I learned are of course relevant to how I deal with my mother. I have to apply them first and foremost in my interaction with her. If I can't do that, it's worthless. Walk your talk. How can I train people if I myself am unable to communicate with my own mother? I often used (mental) tapping when I was with my mother. That way I was able to let her finish talking and I was no longer constantly interrupting her. As a result, I started listening to her better and I started talking to her more calmly. And sometimes, we began to understand each other a little. By letting go of my own pain and judgements, I started to look at my mother and listen to her much better.

The strongest

My mum was in Spain recently, with my sister-in-law and two of her friends. My mother is 84 and those three women took good care of her. It was my mum's fervent wish to go to Spain one more time. But she hesitated, because she was a little nervous. She really wanted to go into that warm sea, but she was hit by a neighbour's car a few years ago. It ran over her leg and the driver just drove on! That fall also broke her hip. After surgery, rehabilitation, lots of training and physiotherapy, she is walking again. But the trauma really shook up my ever resilient mother.

Her fellow holidaymakers asked her time and again if she fancied going in the sea. She really wanted to, and eventually she let herself be helped into the sea, walking between the two younger women. She loved it. But she said she was thinking: "All those people on the beach are looking at me and thinking: Look at that old woman going into the sea. She can't even do it on her own. Why is she even going in the sea? Showing off." I said to her, "Mum, you are not that important. Those people are all too preoccupied with themselves. Whether their bikini is sitting right or their hair looks ok. "Yes, that's just the way I think," she said. Indeed, my mother lies on the beach judging everyone, one by one. She's always done that. We all do it sometimes. Me too. But imagine if you thought yourself more important than all those around you? How good would that be? And it would save so much hassle inside your head.

My mother often described in detail how her older sister used to bully her. She scared my mother and not just when they were little.

Later too, my aunt was quite a dominant presence in my mother's life. That aunt was herself often beaten by my grandfather as a child, with his belt. And, as a young child, grandpa was also beaten and abused by his father. When he was 14, he climbed on a fairground truck and hitched a ride from the East to the South and ended in the West of the country. He never went back. He met my grandmother in The Hague and stayed here for the rest of his life.

My mother never retaliated when she was bullied, even though she was physically strong. She has survived all the bullies and she is still here. She has room in her big heart for everyone. Despite her strong judgements and convictions, she is optimistic and determined. Which helps her get through everything and come out the other side. She's never alone. Many people share their stories with her. She sits in the garden with the gate wide open, so walking past without greeting her is almost impossible. Neighbours walking their dog come and sit with her and are offered coffee or something else, depending on what they fancy and what time of day it is.

My mother didn't realise how strong she was. And after divorcing my father, she brought the next bully into our house. She did so out of fear of my father, who had tried to kill her when he was very drunk. He hadn't succeeded because we fought back. She fought like a tiger and we, her three children, fought alongside her. And we won. We sang songs about that later. And so, we tried to sing this traumatic incident away, laugh it off and hide it away. We never quite succeeded. We cried, laughed, sang and drank. We used anything and everything and did silly things. We were well sick of it.

My mother went from one abuser to another. But this had been her situation since childhood. And she had survived by not retaliating. She said she didn't, because there were three small children in the house. My mother did everything she could to avoid abuse and disagreements. She skirted around the matter and avoided answering my questions. That bothered me because I didn't believe her. I tried to break down that wall and didn't understand why she didn't act 'normal'. But that was her normal. Her survival mechanism. And that's how she had always coped and is still coping.

My mother has often said, "I only started loving my father when I saw him with my own children. He enjoyed being with you all and was always carrying you around." And I remember that. He was the safest man in my life. I knew I could always turn to my grandfather and if I needed to, I would run to grandma and grandpa. He always stood up for me, even if he didn't necessarily agree with me. He once told my sister, "If she'd been my daughter, I would have broken both her legs." But he let me in and I was allowed to sleep on the sofa when he knew I had nowhere else to go.

Survivors are tough

My other grandfather was a terrible bully. And that had affected my father, my aunt and my grandmother. And us, his only three grandchildren. He called us 'ugly rabbit faces'. Who says something like that to children? Or he would ask, "So, how's your old Mum?" Or he'd say weird things about my other grandma and grandpa. He looked at us like we were from another planet. And maybe we were? If he felt like it, he would find your weak spot and poke it. That worked pretty well with my father, so they often argued. Something would happen and we wouldn't see grandma and grandpa for several years. Father and son were seen fighting each other in the backyard until well into old age. My grandpa would be in his 70s and my father about 50. "Oh the shame of it!" my grandmother used to say, drawing the curtains so she no longer had to watch it.

As a child, I felt it all, but didn't understand how unhealthy it was. We used to secretly laugh at him. He asked for that, didn't he? I liked my grandmother and she also joined in with making fun of grandpa. They lived near the beach on a smallholding with lots of animals and plenty of outdoor playing space. So, despite the weird grandpa, I liked going there. The place was big enough to avoid him. We knew in which corner of those beautiful gardens he liked sitting with his miserable thoughts for company, a crate of beer at arm's length.

Grandpa came from a family of ten and out and about, I recognised his brothers by their blue eyes but also by their cheeks, tinged blue from rosacea. They didn't have it easy. They'd probably never even heard of looking inward and finding their inner emotions. As a little girl, I sometimes bumped into my great-grandfather and would always say, "Hello grandpa!" He was stone deaf and I knew that and he had no idea who I was. If I remember correctly, he almost made it to 100 and Grandma (my great-grandmother) was over 100 when she died. My grandmother also lived to be 95. Tough people.

When I think of my ancestors, I can only conclude that they were survivors. They had lived through two wars and were 'enjoying' the country's reconstruction when I was born. But they seemed to be clueless about who they really were or what to do with their lives. And what to do with your children and grandchildren. And that was my 'normal'.

How to be yourself?

Wherever you are born is your 'normal'. Even if it is messed up, unhealthy or bad. It's the only thing you know during the first years of your life, so that's your normal and that's the way it is. The way it should be, the way we do it. And so that's how you do it too, even if you feel that something is not right with who you are yourself, with what you actually want. Even if you feel that you are different, feel different and want different things. If you want to survive here - and you want to, because you have to - then you do the same. So, you adapt. You automatically adopt standards and values and rules of behaviour. Totally subconsciously, because this all happens long before you are even able to think. It is hypnosis and programming, of which you have absolutely no awareness.

I used to really believe that I knew better. With hindsight, that turns out to be not the case at all. When I started raising my child, I sometimes heard myself talking the same old nonsense. The nonsense I'd had to listen to myself. For example, I sang one of those annoying stupid songs because it just popped into my head and it needed to get out. And as I sang it, it really irritated me. That happens quite often with my own annoying behaviour. I can see that I'm doing it but I do it regardless. At the same time, I apologised for it. My daughter is a wise person and healthier than her mother, so she usually just shrugged. Or laughed at my silliness.

Fortunately, I have not repeated much of my family's nonsensical behaviour. For instance, I have never called a child 'rabbit face'. Or poked fun at a child's parents or grandparents. Nor have I ever willfully broken anything of theirs in their presence or even their absence for that matter. I did not drink myself stupid before starting an argument with others. Or get into fist fights with them. I've never done anything idiotic like that. But I do catch myself doing little things I don't want to do. And sometimes I am like the person I don't want to be.

Accept where you come from

The stronger you believe you are not like this, the more likely you are to become exactly the same. The harder you fight becoming like this, the more chance there is that you already are just like it. Accept who you are and where you came from. Maybe we chose to be here now and even to be born into this family? I like to believe this is the case because it feels powerful. I prefer to live with the idea that I have chosen something. Rather than being dumped down here accidentally by a tired stork. Because it was cold and it wanted to get to its nest, so it just tipped me out of that nappy with the van Uffelen family. It was the last day of November, when the weather in the Netherlands is always dreadful. So yes, I have every sympathy for that stork but it is not my story.

And I know the limiting beliefs about myself are not true. Yet too often I still behave according to what I believe somewhere deep inside about myself. And as a result, I keep myself small. And that is a sinful waste of everything I have inside of me and that is not what I deserve. I deserve to live more and more easily, driven by my potential. I'm always full of ideas, I'm creative, entrepreneurial and sincere and that's how I want to live. Just easier, without the bullshit I tell myself about myself.

In a way, I have always known that the ultimate answer to all my questions lies within me. It is not in this body in this house on this street in this city in this country on this planet. It is there again and again as I retreat and come to silence. And yet again and again I am directed to live according to judgements which are not mine. At least, that's what I think. Yet, more often than not, I am surviving, rather than living according to my deeper knowing. Time and again, I care about what others might think of me. And I don't even know. Because I care about what I think others think. Because when I ask others, it is usually not at all what I thought.

So while writing this book, I asked for reactions from 'the others'. To experience how that makes me feel. I sent my first book to the publisher and after a few days, it came back approved. I expected red marks and corrections everywhere, but there were none. It was just right. Maybe it doesn't appeal to everyone, but apparently, it's good enough. Who is the harshest critic here? Is it the others or am I just doing it myself?

Reflecting and distinguishing between my own story and that of others. Tuning into the true me! Whoever or whatever that may be. What is the story of my family and where I come from? And how did we survive? And what have I concluded about myself? What are my own judgements and prejudices about me? And about them?

Ultimately, we are all one unit, but within it, each of us is unique. With our own style and our own path. Original and different from the others.

What is hypnosis?

Kim Jewell:

Hypnosis is when you are not in the present moment. The only thing that is real, the only thing I have power over, is the present moment. If you don't direct the thinking it goes back to the past. It picks up a programme from the past and starts executing it. If you are not in the past and you are not in the present, then thinking goes into the future. And tries to imagine what the future will be. The problem is that we cannot change the past and we cannot predict the future. So the only place for our power is in the present moment. When you are not here in the present, you are in a form of trance or hypnosis.

I often say to my clients: "Tell me exactly what you were thinking when you woke up this morning and you put your foot on the floor." Most people can not remember that, because it has become so habitualized and so trance-like that they go through these motions. They do it so many times that the mind goes: "Oh we can just go on autopilot and not be present" to those menial tasks as the mind would consider it. And so for me it's really important for people to understand that we're not saying they don't want to be present but they have gotten habitualized out of being present and they just kind of sleepwalk through life for lack of a better way of saying it.

This possibly starts with our earliest training. Our parents were busy, they had to go to work, they had to get us ready. So we also learnt routines and no one actually required us to be present. I don't know how that went for you, but no one ever explained to me what it was like to actually be there, to be in the here-and-now, when I was growing up. And as we enter our school years, we also do the same thing over and over again. We get used to it.

And when we have something that upsets us, that makes us feel like it's life-threatening or we experience anxiety, the mental safety mechanism kicks in and it says, "Don't worry about that. Distract yourself. I will run a programme for you. You don't have to pay attention to it." And this creates a snowball effect. This is why you often hear people in their mid-30s and early 40s saying, "Is this it? I don't really feel like I'm living my best life. This is not what I signed up for."

Rob de Groof:

Hypnosis is a tool to adjust programming on a person's hard drive. The state of being in hypnosis feels different for every person. Being in hypnosis is that you are listening to someone, with your eyes closed. And in that moment, that person has access to your subconscious. The critical mind takes a backseat for a moment and then the hypnotist, with whatever technique he uses, can start changing your subconscious programming in order to bring about a positive change.

Ina Oostrom:

Dave Elman says: "It's getting past the critical mind, allowing acceptable suggestions to be incorporated." Many people think that during hypnosis, another person can control them. This is absolutely not the case. A suggestion has to be acceptable and only then does that suggestion pass your critical mind. However, I would say that everything we see in the world, around us, from the moment we are born, actually even earlier, is hypnosis. Hypnosis is another person giving you a suggestion with the intention that you follow it. That's what advertising, politics, teachers and parents do. It's everywhere in our lives. What we actually do is remove from the system all the glitches, which were recorded by often unintentionally negative suggestions. I prefer to call it de-hypnotising. Those glitches start rooting around and lead to emotional inflammation. And you can remove them with hypnosis.

What I love about it is that you help people become their more authentic selves. Because the question is: to what extent are we ourselves? Because of everything we have learned, because of all the structures, education and the way we have learned to think? The best thing to work on is self-confidence and self-worth. Because if that is more present, then people need less from their environment. And then you are less likely to have arguments, because you no longer get angry at each other, as you have much more space for yourself. As a result, you no longer need confirmation from the outside world.

In other words, hypnosis is a shift from your analytical thinking to your emotional thinking. If you have a task to do and it just doesn't work out and you start procrastinating more and more, you are too much in analytical thinking. That's a real problem for many people. But if you are in your feelings, in your flow, you can finish a task in half an hour, rather than hours. Because in that moment, you don't work against yourself but you are actually in the flow of hypnosis.

Martijn Groenendal:

Hypnosis is a technique to communicate with your subconscious. Our subconscious is that part of us that controls all automatic thoughts, feelings and behaviours. This is a hellish system but it can also be a heavenly one. Depending on how you deal with it and what programming your autopilot contains.

How about you?

To let go of what you don't want anymore in your life you have to become completely silent to start seeing the difference between mine and thine.

7. What is my parents' story?

8. What about the other educators who had a (big) role in my life?

9. What becomes possible when I let go of those old stories?

When you let go of everything, you no longer need to survive, you make space for your story. And for your possibilities. Maybe even for something you never thought possible.

Jules expressed and displayed his greatest talent, yet he could not enjoy it enough. Because he kept repeating the voices from the past as he remembered them. He remained the child who had not been seen. And that's how he treated himself. Until he could finally let it go.

BONUS 1

Whilst reading, you may well recognise some of the patterns of survivors, such as having strong judgements and a lot of criticism, especially towards yourself. Are you surviving, considering everyone around you? Then check out this bonus material.[1]

---

1) tapandletgo.com/bonus1

4. Faith, hope and love

"Do you come from a dogmatic background?" I was asked once. "We were nothing," I said. And that is also a dogma. If you are told to go to church, that's no fun, but if you're not allowed to believe in anything, that's no fun either. I almost ended up in a cult once, but before it got too serious, I felt something was not right. Some of my family members got caught up in a highly dogmatic group. Understandable, because dysfunctional families attract and find each other in such 'new families'. There, daily hypnosis is used by the leaders, but the brothers and sisters are not allowed to engage in it. They are told it is very dangerous. Since I got personally involved in this, I have been studying the phenomenon of religion, cult, programming, brainwashing and mind control. And that is also very useful in my daily practice.

I once asked my father if he really didn't believe in anything. He told me that once, on his way from the farm where my grandparents lived to his home in the city, his bicycle got a puncture. He had to walk all the way because that long, quiet road was deserted. In fact, there were houses all along the road, but ringing the bell and asking for help did not occur to him. It was also starting to thunder like crazy. "It came down in buckets. I was soaked through. Then I looked up and I said, 'Is this for real?' Then came one huge bolt of lightning and the rain stopped and the sun came through. It made me think: Am I being listened to after all?" My father's anti-church dogma came from an experience, which I never asked about, but his resistance was strong and absolute.

Our father

When I used to talk to my sister about my father, I would always say "our father, who is not yet in heaven". Shortly after his death, I used to talk about "our father, who is now hopefully in heaven ". And in recent years, I say it correctly. And of course I know that prayer is not about my father, but about that other one. That father of all of us, in whom we believe so much.

A long time ago, I asked my mother if she knew the Our Father prayer. And if she would teach it to me. She wrote it out for me. I wanted to believe in something when I was about nine years old. Or maybe I just wanted to belong to something. To something you went to with your family. I wanted to sing and ended up in a choir. Not for very long, because I felt I was too different. Only much later did I understand that it was the reformed church that organised a club, a summer camp, a Christmas play and some singing experience for the children from my street. "Thank you for this new morning" we sang every day in those summer camps. I have good memories of it and also less good ones.

We can be indoctrinated and manipulated. It happens everywhere, not just in religious institutions. My knowledge of hypnosis is useful in working with people who are indoctrinated. We all are. How about the daily news? Or films, series, books? It all puts you in a trance. We escape from the here and now for a moment. We can choose to do that. But often we don't realise it's happening. Or how it happens. And we have no idea of the distinction between fact and fiction. Does that even exist?

You can leave some religious communities when you've had enough. But there are groups, where the consequences of leaving are severe. You are excluded and you lose everything and everyone. With all the loss, grief and pain, you also get a few kilos of guilt.

Guilt without guilt

During the first session, I help Sigrid (45) to let go. She finds it all terrifying. It's the tip of the iceberg, she says. Shortly afterwards, we agree that she will do an individual retreat with me. Three weeks later, I am at her house and we do a session every day for a week, after she gets home from work. Then we go swimming and hiking together. Her problem is massive insecurity and constant doubt in response to trauma.

As a teenager, she was shunned from the religious community she grew up in. Her parents divorced, partly because of religion. And contact with both parents is difficult. She is full of blame towards her father, who abandoned the family. And she is no longer allowed to visit her mother because she has been "excluded". Contact with her sisters is difficult because they are still "in the faith".

When Sigrid was 17, she was told to appear in front of the elders. She was questioned about her sexual activities, which were non-existent at that age. But the memories of those interviews are painful. The questions asked by some of the older men were inappropriate and crossed the line. During the process, Sigrid was portrayed as a wicked young woman. All this created such resistance in her to the extent that she could not continue in that church. But that meant her mother and sisters had to break all contact with her. Sigrid was blamed by everyone for this break-up whilst all she had done was occasionally thinking about kissing and stuff. She escaped into a marriage she was too young for. It didn't last very long.

Revelations

This led to even more guilt, because this was another thing she'd failed at. She was just drifting and had no idea how to shape her life, something she had never learned to do. Her mother didn't know that much either, which is perfect for ending up in such a cult. Nice and easy under the pretence of so-called love. Where you are told exactly who you should be and especially what is not allowed and what is not possible. Surrounded by brothers and sisters who are equally clueless and keep up appearances whilst also lying awake, consumed by fear and guilt.

The greater the oppression, the more drinking and eating there is. A fairly young elder once told me that he had learned to drink at Bethel. That's the headquarters of his church in New York, where he took a course for elders. For years he had looked forward to being there. This revelation turned out to be a very painful one. He now has a YouTube channel, where he informs people about cult life and where he gives 'shunned' people space to tell their stories.

I also work with religion-traumatised people. They come from different churches and sects and know how to find me, even though I never mentioned these issues on my websites. I have worked with religiously oppressed people, not only when I worked with addicts in the US, but also here in the Netherlands. Even in my private practice, I worked with someone, who had to run for her life as a 15-year-old. Hunted down by her 'brothers and sisters' including her own mother. Because as a girl, she was told she'd become the cult leader's next bride. However beautiful the Bible studies, services and conferences are, the obsession with sex prevails worldwide. And the interference of elders in family life is bizarre.

Sigrid did not manage to adjust. She refused to. She sensed very clearly that they wanted to break her spirit, even though she could not articulate it at the time. Over the years she had found her own way to survive, but the guilt was still enormous. During the sessions, I helped her experience the conversations with the elders one more time and then let it go. Her grief was released, but so was her anger. And her strength. She now saw the elders' frustration and let go of what wasn't her issue in the first place. Whatever reason those men had for humiliating and belittling her in that way, it was not her issue. She was able to let go, heal it and change her perception of herself. She had been an innocent child and the things that were said about her were not true. I gave her alternatives, which usually come to me on the spot.

Only you yourself know God's words to you

We talked about the difference between a bunch of elders and God himself, whom of course we all know personally really well. Religious leaders are people with their own judgements and motives. How can they tell you what God thinks of you? And to what extent you may or may not connect with Jesus? We all have the right and freedom to choose our own faith. And you get to decide whether you believe in Jesus, Mary, Father Christmas, God, the Easter Bunny or that scary bloke from the WHO.

Sigrid said she didn't know how to thank me for helping her restore her faith in life, in Jesus Christ and especially in herself. I don't care what you believe, as long as it works for you. Sometimes I see people deny themselves a lot of things, based on religious dogma. But if someone is comfortable with that, who am I to think otherwise? Sigrid said she felt safe with me because I knew a lot about this particular church and its customs. Funnily enough, there is often loyalty to the church alongside sadness and anger. This includes the brothers and sisters who shunned you and who stop greeting you in the street. Because they believe they are doing the right thing. She did not have to defend them to me and that gave space for her own perception and judgement of what had happened.

Often, social control and fear of criticism play a bigger role than religious conviction. This not only happens in cults, but in any group. And laziness. Not thinking too much, because you want to belong somewhere. 'So don't rock the boat'. And then there is the threat of Armageddon or whatever doomsday that has been programmed into them on a daily basis. Every day for years. That sits really deep. Sigrid felt I understood that and that I neither condemned her nor her church. As a result, she experienced space to open her heart and tell the whole story.

And I helped her clean up and free herself. We also did many rounds of her forgiving herself. To shake off those mountains of guilt and blame. Lots of great releasing. She cried and we laughed. There's nothing as liberating as laughing out loud in the middle of crying. And she wrote to me: "In one week, my life has changed significantly. Thank you so much!" I went back a few more times to work with her, her sister, her partner and other family and friends. It was a wonderful experience to be able to work with members of the same family. An honour to gain so much trust and achieve such beautiful results. All this took place in beautiful Scandinavia where I was also thoroughly spoiled by Sigrid and her sisters and friends. Because I had made such a difference to their lives.

Bells are ringing and angels singing

Joost was someone who saw music in me when I was 15. He taught me to play percussion instruments and how to read rhythm from sheet music. I loved it. He studied at the conservatoire and said I had the right build to play the bassoon. And he asked if I fancied it, because he really thought it would suit me. I was curious and eager to learn, even though I had no idea what a bassoon was. He put me in touch with Lina, a bassoon player from Alaska.

He also arranged for me to get a bassoon on loan from the conservatoire. Back then, such an instrument cost around 7,000 guilders. I was really very lucky and only paid the insurance of 22 guilders a year. But at home, they really did not get it and mostly joked about its sound. "Raise the bridge, because a steamboat is coming." Curiously, we had a bridge in the house to reach the storage cupboard and the dovecote above the stairwell. Who was the crazy one exactly? A child learning to play a particularly fine wooden instrument? Or a grown man, luring pigeons into a hutch?

Lina also taught me to read music and she did so with much love and patience. I loved going to see her and I enjoyed the sounds I coaxed from the instrument. I loved the bassoon but I had a hard time concentrating. Things were difficult at school and I no longer felt at home in my own home anymore. My parents' divorce had not exactly improved things. On the contrary. My sister had married out of sheer desperation and no longer lived with us. And then when I tried to learn something new, it was ridiculed by crazy adults. This was my 'normal'. After two years, the conservatoire needed the instrument back for a student. With regret, I said goodbye. Renting or buying a bassoon was out of the question. There was no money for that of course.

Later I bought a piano and took piano lessons. I have had a piano at home for many years, but I never seriously pursued learning to play. No excuse. An observation. I play it occasionally and for years, I used it to rehearse my singing parts. Because I have always sung. I still do, as did my grandmother.

Willem, whom I lived with a long time ago, noticed that I have a beautiful voice and thought I should do something with it. He arranged for me to have singing lessons. For about 18 years I sang in a really good choir and during those years I learned to read music. Now I can sight read reasonably well, but never without the feeling that I'm not really up to the job. In making music, I suffered a lot from the music culture and dogma at home. In our 'normal' no real music is played. We do sing, preferably with the wrong lyrics to most songs and I also learnt a lot of that nonsense. I loathed sentimental songs, which were listened to and sung at my house. They made my skin crawl.

My grandmother (mother's mother) often sang and from her I learned a lot of old songs. Sometimes, I can hear myself sing a whole song, which I didn't even remember. With grandma I sang a lot. Louis Davids sang: "If you are born for a dime, you will never reach a quarter. Whether you know Greek, Latin or 20 languages, rest assured, life defies you. You imagine you are pulling the strings, but hey life throws you back and forth. If you are born for a dime you will never reach a penny." Typed from memory here, without googling. These lyrics, including the other verses, were hammered in me. Grandma sang this with passion and joy.

Should I continue or do you see how I am programmed to keep myself small? This song by Louis Davids contains no encouragement to become an entrepreneur, to design your own work over and over again, to write books or to address an audience. Or to think of starting up an international business. This was not encouraged, because they had no idea this was possible for people like us. But fortunately, we also sang with Jozef Schmidt "Heut ist der schönsten Tag in meinem Leben." Good for my musical sense, because my grandmother certainly had that.

Cocky and anti-social

I am stubborn, opinionated and anti-social. I don't mean anti-social in the sense of throwing the rubbish bag next to the container. But more in the sense that I don't have much with being sociable. In the sense of turning up at birthday parties and then 'socialising' with everyone. I get bored pretty quickly and it generally doesn't interest me at all. In a group, where everyone can hear me, I don't usually get to ask the questions that I find really worthwhile. I want the best for my fellow human beings, but sometimes my own impatience drives me crazy. Or the lack of genuine interest in who we are and where we come from. And the unwillingness to think a little deeper.

How can people just keep going to church in this day and age, when everyone has heard about child abuse by now? Surely that should set off alarm bells? How can you close your eyes to that, fold your hands and listen to the sermon? Mumbling men in long dresses, wading down the aisle on slippers, surrounded by little boys, also in dresses? Don't ask me why and how this happens. For years I sang in the most beautiful churches with the most wonderful acoustics. But the more I understood Latin, the less comfortable it felt.

I have a thing about Greece and icons. So I have also admired many Orthodox churches of all shapes and sizes. And wherever I found an empty church I sang and enjoyed the acoustics. But there is still something strange about churches and they have been empty for years for a reason. It is magic, enchantment and hypnotism but not always of the purest kind. And because I was not brought up with the church, I used to think it was just me who never felt quite at home there. I loved the singing experience and I loved my conductor. But when he quit and I worked abroad more and more, I decided to leave. And for the past two years, I have been singing mostly with my little nephew and solo at home and on my bicycle.

So that's me, someone who can't fake it. Not that I necessarily think you should always be honest and serious in everything you say and do. I love an entertaining story and it doesn't have to be true at all. I love that and I can really appreciate a good joke. As long as it's not at anyone's expense. But I can't ignore things that really affect me. And there are quite a few of them. That's also why I like being alone. Because it saves me a lot of drivel that doesn't interest me at all. It is the way I am and I have sometimes wished I was able to lie or pretend a little better.

That inability to lie. I found it a huge relief when I discovered I could just answer "Good" to the question "How are you?". Or nothing, but throw back the same sentence. I learnt that from the Greeks and the Irish. If you ask those "How are you?" they just respond with "How are you?". Great! That saves so much energy and effort and time. Matter of perception. I could either give details of what was going on inside me (I thought for a long time that was the purpose of that question). Or I could not answer at all. And still be friendly, because I have at times come across as being rather blunt. Pure awkwardness, but yes, nobody knew that. Neither did I.

I am different. Who isn't?

"When Jeannette wants to do something she just does it." wrote my mother years ago in a Saint Nicholas poem. "Different from your sister and wilder than your brother. You often tore your clothes, but you didn't care. You are different."

My mother recently said, "I have three very lovely children and all three are totally different. Do you know that a lot of people think I only have two children? And then I always ask which ones they know and they mention your sister and your brother. Crazy isn't it?" I think that's because I left the neighbourhood at a young age. And even though I have lived quite close for more than 20 years, I walk or cycle in and out of that neighbourhood, but I am not part of it. I am different. I've always heard that and I think I knew that myself. I never really minded that. But everyone is different, right? Even identical twins are different from each other. I wanted to leave, because I thought it was a small world. I have never worked in my old neighbourhood and have no social life there, except that my family lives there. I really wanted to leave, but I also like being back. I am at home here, close to the sea.

I was told at times that I am too independent. I am loyal and am able to love someone very much. I like that I am capable of laying floors, wooden floors, mosaics, and tiles. And that I am able to make anything and everything and that I run an international business, where I do particularly fascinating work. And I raised a child, while also doing other things. I love to learn and enjoy problem solving. And if I don't succeed, I just ask someone else. Why too independent?

How can you ever be too independent or too autonomous. I don't think that even exists. Hans Korteweg once wrote:

"No one has ever done anything to me
and no one can do anything for me."

Hans Korteweg

In my opinion, that is taking ultimate responsibility for your own life. In every relationship and in every situation. I have known Hans (80) for more than 30 years and know that he has been faithful and has been married for a long time. I read his blogs, in which he shares his own soul searching and practical circumstances. I have never heard someone like him being called too independent. Mahatma Gandhi said it slightly differently:

"Nobody can hurt me without my permission."

Mahatma Gandhi

Taking responsibility for (cleaning up) your past, for your life now and for your future. That resonates with me. To lead my life all by myself and not blame anyone when things don't go the way I want. But that doesn't mean I never ask for help.

I have the gift of sensing if someone is trying to hide something and keeps up appearances. And I've never understood why people do that. It's a diversion to real connection. I have learned to stop trying to help people who are not honest with themselves. In my private life, I avoid such people. And in my work, I am learning to get a feeling from the start as to whether or not someone wants to be helped by me. I don't like flogging a dead horse and if it doesn't work, I prefer to say goodbye to such a client. Being honest with yourself can be difficult, because sometimes you don't realise that at all. That applies equally to me.

Early on, I learned to stand on my own two feet and I love that I can do that. I also taught that to my child. And I teach that to my clients too, so that we don't end up in unhealthy dependency. That's not good for anyone. I like it when people stand their ground. Without any fuss. Taking good care of yourself should be everyone's priority. Instead of shortage, you will create overflowing abundance. Even during my sessions, I emphasise the client's responsibility. And that it is their session, where they have the space to get up and stretch their legs, ask questions, et cetera.

Naturally knowing there is enough

Switching back again to that cute little 2½-year-old boy I have here on a weekly basis. He enjoys sharing food. He loves himself and enjoys food. But always, when I approach him, he likes offering me a bite. Pure pleasure! He is growing up believing there is plenty. But he is also a little glutton, so when he sees some tasty morsels, he snatches them right off your plate.

Very different from my father, who, from what I heard and saw of him, often went without as a child. And became a messed-up little boy as a result. First of all, there was hunger and poverty in his younger years when his father was away. His young mother was turned away with her child by her own mother. Because no one wanted to take responsibility for that child. And on top of that came five years of war including a winter of starvation. My father would 'jokingly' pick meat off your plate if you weren't careful. And he didn't share sweets with us, but hid them in a cupboard. We knew how to find them of course and secretly, when he wasn't at home, we would steal them from him. What could be more fun than sharing sweets with your own children? If you can't or won't do this then there is something wrong with you. And with my father all sorts of things were wrong. And that, of course, had everything to do with what he was told and taught about himself.

Two drops of water

I forgave him in my heart and soul, which of course did me good. It allowed me to see more and more how much I resemble my father. This is also something I have often heard throughout my life: "You are exactly like your father." And I, who was bombarded with stories about what was wrong with my father, was not happy to hear that I looked so much like him. But that has passed. It is something that changed because of my fall down the stairs. First by his fall and then by mine. I will say more about this later. I have come to realise that looking like my father is not so bad after all. The neighbours thought he was a kind man. And I also often heard him described as a handsome man, with his beautiful blue eyes. "Ah child, you have your father's eyes. Two drops of water." And my father was good with his hands and creative. "What his eyes see, his hands can make." And so can I.

I inherited a lot from my father and the older I am, the more I see it. And the more I'm able to love my father, the more I can appreciate and harness it in myself. My total potential. Yes, I also inherited craziness from him and still wonder who he really was. He has been dead for years , so I cannot ask him. Yet I always get answers when I ask such questions. For that, I have to go inside myself. When I start writing or just sit down, I get answers to everything. Is it coming from me or from my father? Does it matter if I get answers? No, it doesn't, actually. An answer is an answer and if it resonates with me, it's good enough. Whether it comes from my dead father, 'who art in heaven' or from the little guy. Or from my child or from the neighbour or from a book. Or from YouTube.You know what I mean.

So I am a child of my parents, who gave me everything they had. Except the English liquorice, which we had to steal from my father. But they gave everything, including all their love, creativity, talents and craziness. Because they were stark raving mad. And so was I. And so are you, especially if you think you're not. If you think you are normal or aspire to be normal, then you are definitely crazy.

Everyone is different. Thank goodness!

Beliefs about yourself are based on what has been instilled in you. It says all kinds of things about your environment and nothing about you. Until you make it your own and become equally convinced. And that happens from the time you are born until you let it go. Survivors teach their children to survive. No matter how. And if you are not so sure about yourself, you can compensate for that by judging the people around you. That can also be done in silence. It accomplishes nothing, but it gives a sense of importance after all.

A client of mine claimed that everyone around her was a narcissist, including her own child. Huh? "What about you?" I asked. That was not what she wanted to hear. Where did I get the nerve? But if you only gather narcissists around you and bring one into the world yourself? Isn't it time to consider this ? So what am I doing that I've become such a narcissist magnet? Because now she was spending entire days blaming them for what they had done to her. She had a whole family full of narcissists. But unfortunately, you can't change those narcissists. You can, however, do some finding out within yourself, release stuff and change your perceptions. How did my own child become such a narcissist? Who raised him? Oh yes, I did. Oops.

When you have become lost in victimhood, it is hard to let go. Especially when you have learnt to survive like this and see no other way out. When you have clung to yourself like that. With survivors, there is often a huge lack of love. They have not experienced it because they had to survive and there was no time for love. That doesn't mean there was no love. Love sometimes looks weird. "Finish your plate , because during the winter of starvation..." Or: "You are not allowed to do this now, because when your brother ... "

People who lose themselves in victimhood also tend to constantly give themselves a hard time. The negative self-talk. "Don't be so stupid all the time. Please stop doing that. Just look at it." This can be constant and sometimes it happens out loud and even involves asking loved ones for affirmation. "See how stupid I am being again?"

Terrible for the person themselves and also for those around them. Because you are required to stand up to it, but at the same time it is not accepted. This behaviour, as I mentioned earlier, is clearly symptomatic of traumatic experiences. But that is the last thing the person in question wants to deal with.

How do you know if you have experienced trauma?

At the end of each conversation with the specialists, I asked if they wanted to add anything of their own. Kim Jewell said the following:

"If people are wondering whether they are traumatised or not, they might want to pay some attention to their inner world and dialogue with themselves. If you find that you are very critical and often busy in your head, that is definitely a sign that you are living in survival mode and that there is something your thinking is trying to protect you from."

You'd rather keep yourself busy than allow silences into your daily life. You prefer to pile up more problems rather than be aware of what is going on below the surface inside yourself. You don't do this consciously and never on purpose. It has become a habit and you have come to believe that this is just the way you are. Or that it is because of your ADHD or whatever label you have been given. But it has become your way of survival.

"If you find yourself often calling yourself "stupid" or saying to yourself "Why do you always act so stupid?" Or "Why did you say that?" That kind of talk tells you that there is something you are holding onto for yourself. And that needs to be erased."

Kim says here that you are holding onto it for yourself. You don't even realise that you are doing it because you are so used to it. You are constantly listening to this negative commentary. And imagine what your life would be like if you stopped doing that. Or if you started addressing yourself in a more constructive and loving manner . That is exactly the change you can make for yourself. It is nice if there is someone to point it out to you. Even if it is that irritating colleague. See it as help rather than criticism. It's trickier if no one hears what you do because you don't say it out loud. Then no one can point it out to you. Hence this advice: start listening carefully to how you address yourself.

How about you?

10. How do you feel about yourself now, about how you look and what you do?

11. Where do these beliefs come from?

12. And is it really true what you have come to believe about yourself?

Sit in front of the mirror and take a deep look at yourself. Yes, crazy exercise, but just do it. Don't look at that hair or those wrinkles or that nose, but look deep into your eyes. Who do you see? Say something kind to your reflection. Whatever you would say to someone else from time to time. Do it. Because it's unlikely someone will be ringing your doorbell today to tell you how wonderful you really are. So tell yourself. Because it cheers us up. But you have to mean it and do it wholeheartedly. Does it not work straight away?

It is a matter of practice. Every day. Say something like "I love you" or something like that. Or "How beautiful you are!" Just make it up and DO IT!

5. Fooled

A seemingly brief moment of inattention is over in no time. But it can create a major emotional problem in a child's life. The child may draw a conclusion, which may not be correct, but which it experiences as such. In Jente's story, her mother's reaction said everything about mother's inability at that moment. But the 10-year-old girl had no idea of her mother's pain, so she related it to herself. And this was responded to from the superficial view of an expert. And so the problem only got worse.

Anorexia nervosa is called a mental illness by psychologists and psychiatrists. I prefer to see it as a result of unprocessed experiences. A survival mechanism that is learned and can also be unlearned.

Jente (15 years old) came to me via her mother. Her mother had done a few sessions with me and had recommended them to both her daughters. First, her older sister had been and then Jente wanted to try it too, because she had a goal in mind.

"What is your problem?" I asked her, after noting down some of her details. "In other words: What do you want to change today?"

Jente would soon run a marathon to improve her personal best. She ran six days a week, sometimes seven. Because she loved running, but by now, it had also become an obsession. And although she was very fit, she was often too tired, too thin and too preoccupied with food. Last time she had failed to beat her record, she had been feeling down for weeks. And she was afraid this would happen again, but not competing was really not an option. "Running is all I am" she said.

She had been diagnosed with Anorexia Nervosa. And OCD (Obsessive Compulsive Disorder), compulsive thoughts and actions. When she was 12 years old, her parents no longer knew what to do with her. They were afraid she would starve herself. So they had her admitted to a special clinic. This was on the advice of a child psychologist, who could not offer a solution either. For three and a half months, she was forced to eat. She felt lost and abandoned. "Every day was the same and took an eternity."

I don't matter

After gaining a few kilos, she was allowed to return home. At home, Jente became obsessed with eating and especially not eating. Her experience in the clinic was not talked about. In running, she found an outlet. She started training with her father. He is a policeman who likes to challenge himself and who is in good shape. Everyone praised her for her performance. People thought she was doing better until her running became compulsive.

"What makes you happy?" is also a standard question during my procedure. Running is what makes Jente happy. She enjoys her body and her strong legs. In the forest environment, she appreciates the space and recharges herself through deep breathing. She loves pushing her limits. Chilling with her girlfriends also makes her happy. Or watching films with them, listening to music and dancing. But she does that less and less because she always has to train.

In the space of a week, I had three sessions with Jente. During the process, the story appeared. She was actually a very happy child until her parents bought a new house. Just after the move, little Jente (9 years old) felt lost in the big new house. The place was chaotic and everyone was busy creating order. The little girl felt lost sitting there in that huge space. And when she sought safety with her mother, the latter focused entirely on a conversation with the contractor. Mother's attention was totally focused on the perfect state of the house. There was no room for emotions. It had to function and, above all, it had to look good.

Jente's conclusion: I don't matter. I am unimportant. I have no place here. There is no comfort, no security, no space for me. She had totally lost control as well as her safety. And what does a child like Jente do then? Create control! And so obsessive thoughts and compulsions can arise. Something is done that gives rhythm and calm. Even if such actions seem pointless, they serve a purpose. At least in that moment.

Jente's mother is a top entrepreneur and her father has a government job. They are well off. They love each other and their children and they try to give them the very best. Jente's mother told me during her sessions that she had failed as a mother. Actually, she couldn't admit to this and kept up appearances instead, until a friend told her about me. She came because she did not know what to do with Jente anymore. During the third session, she had a breakthrough, resolving old pain from her own childhood.

She was then able to convince Jente to give it a try. Since her admission to the psychiatric clinic, Jente had not got along with her mother at all. She was permanently angry. She was verbally aggressive with her mother. And she had very little trust in therapists.

Pulling out the root

Of course, Jente's story did not appear by itself. But I have a good method to elicit the story connected to pain. Sometimes, it works faster than others, but on average it happens during the first, second or third session. It is similar to pulling out the root and planting a new seed. In the second session, Jente was able to feel her anger and sadness again, before facing it and letting it go.

Despite being fully awake and consciously doing the session, she was also in a trance. This is not unusual. We are all in a trance when we tell a story. Because our mind floats away towards the memory. We see it in front of our eyes again. Or we hear the voices. And we feel again what we felt at the time. Through my questions and interventions, unprocessed emotions are resolved at a subconscious level. This gave Jente a new perspective on the story. Jente told her father that she felt heard for the first time which gave her a better understanding of herself. Before, her father had had his reservations about my work. But he was impressed when he saw the change and calm in his daughter. And, of course, he was especially relieved.

"Educate yourself and leave your child at peace."

Jiddu Krishnamurti

After Jente's three sessions, her mother came for another session. She knew the cause of Jente's problems and still felt regret and guilt. But during this session we also worked on her own pain from childhood. And she understood what created her inability to help. This created space for her failure as a mother. And she was able to forgive herself and let go. Within a week, I did these sessions with the mother and her daughters.

You cannot turn back the past. We all make mistakes. Even with our children. You can face it, forgive yourself and lovingly let it go. You shouldn't look for the cause in your child, but in yourself. And these are not necessarily mistakes. That too is a matter of perception. Do you see disasters, accidents, bad luck or do you see the lessons in them? What you see is how you perceive it. And that is why you don't need anyone to blame you or to forgive you. You can sort that out yourself, inside of you. It means the ultimate power is within yourself.

Jente ran the marathon. She had trained for it for a year and in the last session we worked on emotions around the result. The evening after the race, I asked her via WhatsApp how it had gone. She had not improved her record, but was relaxed about it. Running had been the only thing she appreciated herself for. By now, Jente knew she was more than a runner or an anorexic patient.

89

A month later, I asked again how things were going. For years, Jente wanted nothing to do with her mother, but she also regretted the way in which she'd blamed her mother. After cleaning up consciously and on a subconscious level, there was understanding. By forgiving and allowing love again, they could connect again. They had made peace with themselves and each other.

You can get rid of it

The cause, the start of the OCD and Anorexia Nervosa had nothing to do with food at all. And if you don't know the cause, you won't find the right solution either. So, forcing a girl like that to eat does not make things better. On the contrary. In the clinic, Jente was put under pressure to eat as a 12-year-old. In some way, she understood, as she was too skinny, but once again she felt not listened to. So when she came home, the problem had got worse.

Food was not the solution. We had to find the sadness and give room to the despair she had experienced. Merely talking (at the cognitive logical thinking level) doesn't get you there. Instead, you need to access the subconscious, where everything is stored.

Far too often in my practice, I hear stories of years of cognitive therapy that don't work. A young woman says in a testimonial on my website:

"Jeannette is direct, but then I think: It's better to be honest and get to the core than to bullshit around it. She asked questions, which got to the core and then she started talking and tapping. And that released emotions that had been locked away for years. And it wasn't like going home and feeling like crap again, as in previous therapies. It was really cleansing. And a massive release!"

Anna had therapy in a regular institution for years. In a video testimonial on my website, she literally says:

> "I was always told the same story by mainstream mental health care. "You give it a place and then you can deal with it a bit better." But I really feel now that after two sessions it's gone. And that something very restful and positive has appeared instead. I have a diagnosis of ADHD of the highest order and of the most severe kind. I am now convinced it was a misdiagnosis. I think it had a lot more to do with trauma, which gave me a lot of anxiety and unease. And that just goes away with this method. I often marvel at it and think: How is this possible? Why not 20 years ago? Anyway, I've had mainstream therapy for years. And that got me nowhere. It only brought up crap. And here in your sessions, you also go to that feeling and it's very intense. But I also know that when I leave here, it's completely gone. And then I just feel really good. And I don't have that with the mainstream treatments. They actually just made me feel worse. Now I feel good and I also tap myself. This really works and I'm very grateful for it."

Resolve or learn tricks

Mainstream mental health services sometimes perform poorly and that's not even intentional. Of course, the people working there usually have the best intentions, but they are trained in the same limited perception. I too have spent years tinkering with myself and doing all kinds of education and training. Higher vocational education and therapies which helped me relax for a while. But there, I never learned to fundamentally clean up the mess and to just let it go.

Another client of mine, Heleen, has had 15 years of psychiatric care, both as an outpatient as well as multiple admissions, including a stint for a whole year in a secure ward. When she told her social worker that she was going to work with me, she was concerned. Heleen was asked what exactly this method was and she was duly warned. And she was offered to be admitted the same day. This was her chance as there was room for her. She was told that a bed was already waiting for her. And the Lithium was also ready. They would experiment with it once more to find the right amount. Like Anna, she felt awkward telling her story, as if she was betraying her caregivers. Because there is also loyalty to the counsellor who means well, of course. The therapist, psychiatrist or community social worker doesn't have it easy either. But by now, she had been through a couple of ECT (electro convulsive therapy) treatments which hadn't achieved anything either. And she certainly didn't want that anymore. She said:

"I am like a monkey, constantly learning tricks to deal with this. But I want to get rid of it, even though I don't believe it's possible. And I don't want to explode my cranium again. That was very painful and terrifying, despite the anaesthetic."

I had told her she could get rid of it and she had finally decided to have a few sessions with me. And even during the first session, she experienced profound changes. Intense crying, forgiveness, lots of letting go and deep relaxation. She was impressed and couldn't believe this was really possible. Understandable, as she hadn't heard anything other than what Anna had also been told: "You have to give it a place so you can learn to live with it." No one had come up with the idea that you can also get rid of it.

"And now with you, during the first session, so much has already been cleared and resolved. I really didn't expect that I could ever feel so comfortable again."

Even well-meaning psychiatric nurses assume it cannot be resolved. And others in the profession. And they tell their patients and clients that. Many healthcare workers are not in a great place themselves. They eat or drink too much. Or they are on pills themselves. Having worked in psychiatry and crisis care, I really know these types . I was one myself.

Dare to ask questions about everything

I did not like history class at all. I can still see that book in front of me, my adolescent self hovering above the pages for hours but it just didn't sink in. I had no interest in what I was taught at school. Useless. Imposed education. And I couldn't remember those useless dates either. I had completely different questions. I grew up with a lot of war stories. And asked why that wonderful queen Wilhelmina and her whole family had left the country during the war. Those questions were not appreciated at school. My Surinamese teacher was himself drilled in 'our' history, of course. He had no doubt belted out patriotic songs about the Dutch 'Motherland' while living in the tropics. He thought my questions were funny, but we really didn't have time to deal with them.

But I always ask: How do you know it's true? Who tells the story? Who wrote the history books? What happens to the money I have to pay in taxes? How about the royals? Why do Amalia and her sisters (Dutch princesses) get 1½ million Euros when they turn 18 and my child does not? What is the difference?

What about my family? What do I know about my ancestors? Of my bloodline? Apparently, my surname has French Huguenot origin, but what does that even mean? Didn't we all originate from there? Fleeing from monarchs, tyrants, popes, templars, Rosicrucians and others?

And that history is linked to my family name, my body, DNA and bloodline. But what about the person who lives in my body, let's call it my soul, my spirit for a moment? What do I know deep inside of me? Who am I and why am I here right now, in this body on this planet Earth? Why is there such resistance and reluctance when I want to find out?

How do we know what is true? Our history has been falsified. In my city, street names are named after 'naval heroes'. They travelled on ships and sailed to other continents. Once docked in foreign ports, they massacred or enslaved the population. They looted and plundered, took whatever they could lay their hands on and became obscenely wealthy.

The victors write history. But shouldn't we wonder who they were or are? Amsterdam's canal houses bear witness to this, as do its museums. But those weren't naval heroes, mind you. They were pirates. Now, we would call them narcissists, self-centred idiots, power-hungry people. Yes yes, it's all about perception. I'll repeat it a lot in this book.

"Take what you can and give nothing back."

Captain Jack Sparrow

Are we being fooled?

Have we been fooled for centuries? How do I know what is true? We need to ask that question more often with regard to everything we have been told. And do we have to agree on everything? How free am I allowed to be and when does it bother you?

Quite recently, I needed to update myself with the Basic Medical Knowledge and Basic Psychosocial Knowledge courses. It took me a year and a few thousand euros and yielded extremely little. Very little new under the sun. But what struck me was the increase in psychiatric diagnoses. Ridiculous. Before, there were two eating disorders. Now there are six varieties of obesity and that was just for one of the disorders. Years ago, it was simply called being too fat or too thin. That there are emotional roots won't surprise anyone. But the number of diets that are prescribed in great detail for these conditions, is downright bizarre. And the number of treatments with supporting 'medication' has really got out of hand.

Are we being fooled? Is this organised by the pill pushers? The drug dealers, the pharmaceutical industry that wants to sell their products? Anything that offers a little alternative help is still being backed into a corner. Previously approved drugs are suddenly withdrawn from the market and cheaper home remedies are banned. It is suddenly called dangerous and unsafe. And this has become much worse since the early 2020s.

We can just learn to let go of the underlying emotions. Then we don't have to entrench ourselves into judging ourselves or others. We can find peace within ourselves pretty quickly and not care so much about what others do with their lives. If you are okay with yourself and know in your own heart who you are, and recognise your own worth, then you don't feel like arguing or piracy at all or like stuffing yourself with food when you are not hungry.

When you feel really bad, you want a solution. How free are we and can we really free ourselves? I'm sure we can. Just let it go. Especially our underlying emotions. But yes, you have to feel that's what you want to do. And it seems most of us find that too much trouble. We stopped believing it's possible long ago. And have no idea of who we might actually be. How worthy we are and how valued.

When I used to say I didn't feel like doing something, my mother would say, "Then make yourself feel like it." And if I said I didn't want something, she'd say, "It's not about what you want." So sometimes you just have to make yourself want to do it. Especially when it comes to your own freedom. Especially when your life is dear to you or if your life depends on it.

Falling flat on your face

Sometimes you just get lucky, like I did. I didn't have to make myself examine my underlying emotions when I fell flat out. No willingness, no unwillingness and no desire to. But if you don't look for anything then it just happens anyway. You may break a few bones or something much worse will happen. I was lucky, I really was. But it was also a kind of alarm bell, a wake-up call. What do we call something like that again? It made me think and lessons were learned. But habits set in and if I am not careful I just stumble along. You will find the details of my stumbling wake-up call in the next chapter.

Can using hypnosis resolve trauma?

Ina Oostrom:

Yes indeed. We work a lot with trauma and often it can be resolved in a few sessions. Usually in 1 to 3 sessions. Some people find it quite challenging to let go of analytical thinking. We have lots of techniques to achieve this, but I still have to say to one or two clients a year: "We're not getting anywhere." And then you shouldn't keep going. There are various therapies available to solve similar problems, so people can find a modality that suits them.

Rob de Groof:

I don't often work with trauma, because it often involves a lot of stories and I have other therapists for that. My latest book is called 'Stop it! The Art of Simple Hypnosis'. Based on Bob Newhart's comedy sketch where he says "We don't go there." I want to keep it simple and basic. That, of course, is thanks to my training with Jeffrey Stevens. I am not a fan of 'regression to cause'. I'm not saying you can never use it as a technique, but certainly not every time. I assume that the subconscious KNOWS why the problem was created. If you give the subconscious enough reasons to make those changes (and that's often where the problem is) then you don't have to clean up everything that happened in the past. If we can solve the problem right now through direct hypnosis then the subconscious mind can work it out. And why do I find that much more interesting? Because often, the person doesn't remember the trauma at all. So why open up the cesspit? It can create a problem instead of a solution.

I just did another session with someone, using 'symbology'. That's a technique by Jeffrey Stevens, where the client doesn't even have to know what the problem is that causes the trauma And where that we actually start turning the feeling around. Turning a negative feeling into a positive feeling. Jeffrey used that a lot with veterans in the United States, who came back from their mission and suffered PTSD.

Martijn Groenendal:

Yes, absolutely. Memories that are unresolved can be revised in different ways using hypnosis. Whilst in a hypnotic state, you go back to a memory which can be re-imprinted or changed, using all kinds of different techniques. This in turn can take the emotional charge out of a memory. It can create relief, because the whole meaning of that memory has been reversed. So that instead of trauma, a huge amount of wisdom, insight and lessons are experienced.

If you experienced certain things in your past that were traumatic, then in many situations, those were traumatic for a reason. This may mean that you were missing a particular skill at that time. And in our childhood, this is quite often the case. For example, being able to say "No" and stand up for yourself. If I'm still unable to develop that skill in my adult self then I'm feeding my old trauma and I'm not growing as a person.

Kim Jewell:

The short answer is YES, but not in the way most people think it goes. Because I take my clients on a journey, back in time. In this 'trance journey', they have the opportunity to change their perception about what happened in the moment that was perceived as traumatic at the time. And what is so beautiful about this way of working, which you and I apply, is: they don't have to retell the whole story.

We take them on a journey from present awareness. They have already been through it and they have survived whatever it was. They now have the skills to help the part, which didn't understand what was happening to them at the time. And that's what makes it so effective. It's a gentle method of working. And it's self-empowering, because it's not someone else doing it. We go back there with them and say, "Wait a minute. Let's look at this through a different lens." And so you replace it with other emotions. So we empower our clients, through the lens of their present consciousness, to help that younger part gain the wisdom of now. Of how they have grown, after that experience.

How about you?

13. Are you willing to put in some effort?

14. Do you dare to stand alone?

15. Can you detach yourself from what you are used to or attached to?

Imagine what becomes possible when you let go of what others think about something. True freedom is found in yourself, in being true to yourself and to what's present in your heart.

And no one can take that freedom away from you. It is your birthright. Take some time to reflect on this.

BONUS 2

Have you often heard that you are impatient, short-tempered or just too slow and an eternal procrastinator?

Or absent-minded and forgetful?

Or do you have a serious diagnosis and doubt it? Then check out this bonus.[1]

---

1) tapandletgo.com/bonus2

6. Enlightenment

Abel is a master procrastinator. But when he suddenly feels completely out of it, he calls me. He asks if he can have a session as soon as possible. We talked about this a few weeks ago and I know I can help him. But I am about to fly to Oslo the next day, so when I say I can see him in two weeks' time, I can tell he cannot wait that long. We agree to do a session early the next morning and then he will take me to Schiphol Airport. He has to go that way anyway.

He has great difficulty in ending his relationship. All beginnings are difficult, but once you take the first step, you are on your way. He knows it, but it hurts.

Dark past

A common question in my sessions is "Have you been through this before?" And so we usually quickly get to the links to what the 'normal' is, in which you grew up. Abel paints a picture of the mother he grew up with. This is how he learnt about love and how he loves. Love, dependency and drama belong together in his world. He loves his partner dearly, but even so Abel feels that he has to continue on his own.

He constantly defends her, even when he talks about her aggressive behaviour. He explains it all away and blames himself. Of course he does, because that's what he's used to. But he can't take it anymore. He wants things to be different, yet a large part of him does not believe this is possible. Certainly not for himself. Despite love, there is not enough room for him. They are increasingly restricted by their codependence. He has idealised his partner and for a long time did not want to see how hard it is for him. He is used to drama.

Abel cries and sobs a lot during the first session. Many tears flow and we are making great progress, but he still has to go to work today and I still have to fly, so we have to drive to Schiphol on time. As I finish the session, I know there is still a lot to clean up. He has let go nicely and is deeply relaxed. But also a bit lightheaded, so I give him lots of water to drink and something to eat. Usually, I advise everyone to take some time for themselves after a session and not to dive straight into the next appointment. But needs must and we are glad we did it today anyway.

When I am back in the country, we have a few more sessions. Even though he is still hesitant, he experiences that it works and does him good. He tends to postpone the sessions. He dreads them because he knows we go deep and he's afraid to let go. Abel has had to let go of a lot at a very young age. I take him to memories he has tucked away deep down to survive. He lets go even more and more water flows. The deep worrying wrinkles fade from his kind face. He becomes more and more detached from beliefs, even unaware but that doesn't matter.

The difference is noticeable. The dark cloud that hung over him is disappearing. His behaviour changes. He is still slow, yet he is making progress. He is standing up for himself more and, every day, he's becoming a little more true to himself. For years, he has lodged more than lived in his flat. Now, he wants to make it a home again, his own place, his own nest.

As I told you earlier, I have a specific procedure in which I ask in-depth questions. And so, I ask Abel about his desires for the future.

Enlightened future

As survivors of trauma, we don't realise its impact. We get used to living with limitations and believe in who the survivor is. But we are much more. Abel is a healthy, sprightly man. He is good-looking and seems attractive as a potential father. Especially in the eyes of a fertile woman who is a bit younger than him.

But he does not see it that simply. He has strong judgements and sometimes comes across as downright arrogant. It is part of his survival mechanism. We secretly consider ourselves to be unworthy while at the same time, we overestimate our worth for spurious reasons: Judging and condemning is a hardening of underlying emotions and desires. You will always find like-minded types, with whom you can sit and judge others.

For instance, he has all kinds of judgments about starting a family and about women his age or younger who just want children. It's bourgeois. Anyway, he is attracted to older women, he believes. That turns out not to always be the case, but this was what he was used to, including the drama.

Abel is quite willing to procreate, but he no longer believes this is for him. His partner had children his age, so having children together was never an option. He has not felt this desire for years, but now it is there. During the sessions, we make room for his deepest desires.

I help him remember old pain and to see and feel one more time what his own childhood life was like. Such painful experiences sometimes make us afraid to bring a child into the world. He experienced a lot of loss at a young age. Separation from his father and a little later the death of both his parents and his brother. He gains more understanding and compassion for himself. Letting go of the darker part of his past makes room for a brighter future and for more love, attention and care for his inner child.

Anyway, to make a long story a little longer: a year later, he met the woman who suited him and not long after that he became a father.

You are no longer the same

The past is gone and never comes back. You are not the same as you were then. Every experience does something to you. At best, you learn something from it and that changes you, so the same thing can never happen again. But you can keep repeating the experience in your mind which causes you to create and attract the same thing.

When you consciously look for the memory, relive and release it before replacing it with a new story, you remove old pain and change your perception. There are many ways in which you can do that and I practise a few. The best of these techniques is that you let go of the traumatic memories and replace the old story with your new story. The quickest way to do this is through hypnotic trance. Because then you are less susceptible to logical, reasonable, rational, judgmental thinking. The critical and therefore limited and limiting thinking. The learnt, programmed thinking. The program, which was instilled in us by church, education and upbringing.

I also often think: What does it matter if I am there? Should I say something? Should I try so hard? What am I doing it for? Doesn't everyone just have their own thing to do? I too am replaceable. But apparently I am here with a purpose and that's why I have to complete it. There is a reason I am here and that is why I am writing this book.

The tricky part is that I am also still largely ignorant and unaware of my own limiting beliefs about myself. So I do what I can from where I am now. That's all I can do. And if I don't move forward, life will make sure I somehow move forward anyway. Actually forward? Don't take that too literally.

My own trauma live show

On the 2nd of March 2017, I literally fell into my trauma. I experienced this as terrifying and terribly interesting at the same time. Human beings are my eternally fascinating object of study. And I observed and felt and saw and did and distanced and was in the middle of it all at the same time. I thought "Damnit, what now?" and I thought "Jesus, this is interesting". Yes, I'm sorry, but in such moments I totally fall into the language of where I come from. And there, there was a lot of damnits and Jesuses, swearing and cursing. I thought "Get out of here" and I thought "Jeannette, you have to stay with it. You need yourself now." All sorts of things happened that I'm unable to describe but I'll make an attempt.

Funnily enough, I know I am in shock. I am experiencing trauma and I am aware of it at this moment. I think I am in different layers of consciousness at the same time. In different frequencies. I do not know if and how this is possible, but it is interesting to experience how I am observing myself and what is going on, here and now.

Right now, I am sitting on the stairs and looking at this weird piece of my body, which is part of myself. It is part of my body but looks so strange and unfamiliar in this position. I feel a strong urge to run away from this, from what is happening, from this situation, from this life, from this world. But I can't, I can't get away from it. What I am able to do, and I know I am a master at it, is to let go of this situation in my mind and dissociate myself from it. But that's not very useful and not very smart, because it won't help me now.

I need myself very, very much right now. I have to stay present. I've always been good at taking my emotions out of a situation. Not feeling any pain so that I can react for the benefit of everything and everyone but me. I am the best, the most responsible and reliable person in any crisis. But now, right now, I am unable to dissociate myself from the pain, because it dominates everything.

The pain is so strong and so sharp, my leg looks weird and I have to do something for my own benefit now. And I know I can. I have to ask for help and surrender to this fact. I choose to do just that. I find my

108

phone and tap Elia's number with trembling hands. As I find myself able to ask her to come and pick me up from the stairs, I feel the urge to panic. All the frightening thoughts and emotions race through me, my mind and my heartbeat. What madness.

Shit happens, shift happens

This is new to me and at the same time I know it so well. I feel my body functioning on a different level, on a particular frequency and I know my brain is functioning on a different frequency. I know that this is just my body, this is not what I am. But I am fond of this body, this vessel and I love having two legs that I can walk and run with, experience life, swim in the sea with, move and be independent. I love my legs. I know I don't need them, I know I can live without them. But I love having them. They are my friends, so I decide to take care of my left leg. It is my friend and I am going to heal it, no matter what is wrong. I don't like the foot right now, it doesn't look like mine. The foot I've always had, the foot I could always trust for walking, running, escaping. There is no more time for denial.

This is a dysfunctional foot and that is not the kind of foot I want. But I need it and this one is still attached to my leg. The skin is not broken and it appears to be the same foot. But my foot is at a totally different angle. I need this one. And I know I am making it easier for myself and for the foot down there to become friends again. There is no point in blaming it for lying there. It is now time for problem solving, time to act.

My brain is functioning fast, like everywhere and nowhere at the same time. Crazy, but very sharp, sharper than ever before. Changing, shifting, I notice that something is shifting somewhere. Shit happens, shift happens. So quick, so fast. While this is going on I observe my brain with fascination as well as myself, my behaviour, my thinking, my leg. I am acting in shock and at this moment in complete peace with and acceptance of the situation. My mind is so clear, so sharp. No time for mistakes, no need to be tough anymore, no reason to complain. I am safe, I feel safe. When there is nothing to lose, there is only surrender. Elia comes and picks me up and takes me to the hospital. In great pain, I am able to hop into the car, dangling foot and all. I decide to stop worrying. That doesn't work for a second, but it's a great decision. Now I want the best hospital in town and the best doctor to fix this little problem. And I know it will all be fine.

Time to patch me up

We ride out of the lift and enter a chilling science fiction movie scene. Here the operating theatres are situated in the basements. Smart, because unlike back home, they don't have winters here in Cyprus and a pretty bright sun shines here above ground all year round. The nurse hands me over at the door, where my bed is passed to two Greek gods. And it's all fine by me. This is what I have been waiting for, even though I don't really know what awaits me. What I do know is that it is finally time to patch me up. All day, I have been waiting for a slot in one of the operating theatres. I had to wait 8 hours to digest one Greek yoghurt before they could put me safely under anaesthesia.

With barely appropriate enthusiasm, I greet the beautiful men who - strangely enough - greet me just as enthusiastically, as if we are going to do something super fun. They get busy with hoses and plug me in and on devices as quickly as possible. And then they tell me I can slide from my bed onto the operating table. "How?" They make sure the two surfaces are side by side. This could be painful and also a bit embarrassing. It already is as I have nothing on except a blue gown which is open at the back and too short at the front to cover my private parts and I have to spread my legs to get from the bed to the table.

Shut up and move on

Now, I have been on a bed with a Greek god before when I was also naked. But just completely naked and not wearing one of these silly gowns. Open at the back, but high necked at the front. Who came up with this? In any case, previously, I also had both my legs at my disposal and could open and close them freely. Today, that action is less smooth to say the least. On the other hand, today I decide to be super flexible from now on. Not so much in the legs, but in how I want to live and act in relation to myself and the rest of the world. We already know that I am stubborn, independent and determined. Shut up and move on, either way, embarrassing or not. They are nurses by the way and they drag me from the bed onto the table. One of them tries to adjust my left leg a bit and it sends a surge of pain through my entire body. Stephen appears just in time to save me. "'Αστο" I hear him say to the nurse. "Leave it." He saw me grimace in

111

pain. I guess he's thinking: she'll be under soon, unable to feel a thing and that's when I'll chuck her leg over.

He looks at me and asks how I'm doing. "I'm very happy to see you again, although the circumstances could be better, like in a less creepy room." He agrees with me, he says. But I'm really very glad he's here. He is the traumatologist, i.e. the wound expert. Trauma means wound and logos means knowledge. He is also a foot specialist and surgeon. This morning, he explained exactly what is going on and what he is going to do and I have confidence in him. My ankle is broken in several places so he is going to cut it open and piece it together. All day, I've been getting painkillers. I thought at first that the tapping was working really well, until after a few hours, a nurse came with a new bottle. He plugged it into my IV. Only then did it dawn on me that this wasn't nutrition but dope and that this was what made me so relaxed. Despite this, the smallest movement made me shudder, especially when I looked at my mutilated foot. But now I am cheerfulness itself and really nice to everyone around me. This really isn't like me, because I hate doctors and hospitals. I never go there out of choice.

Nothing to choose

But today there is nothing left to choose and I have decided to go along with that. I realise I need all the help I can get, so I can no longer afford to hate doctors. There is one in front of me now, who has my best interests at heart. And who also doesn't look that bad at all. Elia filled me in: She has known him since high school and wanted to tell me all sorts of juicy details about him, but I didn't want to hear them. Not yet. Maybe later.

He walked into my room this morning and introduced himself. He hung his jacket over the chair and sat down next to my bed. He took all the time he needed to show me the x-rays, explain the procedure and answer my questions. He apologised because I had to wait a few more hours. I didn't care anymore. We had a date.

"Nice!" I say to Stephen, who is tying a pirate bandana on his head. "That's my favourite colour." Months later, I suddenly wondered if he really wore a bandana with little fish on his head or was I imagining this because I was on drugs. Little did I know that the bandana was the latest operating theatre fashion? I hadn't been in an operation theatre for years, so had no idea what they wore there. Only when those bizarre videos of hospital staff dancing in 2020 appeared online did I see that bandanas were the latest fashion. "Thank you!" he says.

The anaesthetist has now arrived, introduces himself, tells me what he is going to do and off I go.

The last anaesthetist at my bedside congratulated me. "On the birth of your beautiful daughter," he said. That was 21 years ago and at this moment she doesn't know I am going under anaesthesia. I was unable to call today to tell anyone at home that I am in a hospital. All day I thought about it, but I also kept thinking, "No, not yet. Too soon. First that foot has to be straight again. Only then can I talk." I had only sent my choir's conductor a message telling him not to count on me for the upcoming concert. "You don't sing with your foot, do

113

you?" he said. "Come home first and we'll talk. Strength, get well and love."

Hours later, I wake up in a bed and find my phone lying next to my head with earbuds attached. I lie on my back and look at the footboard and see that my foot is straight again. All wrapped up in white bandages. I make an effort to move a little and feel no pain. I read a message from Elia, who has sent me a Louise Hay playlist. So I put in the earphones and send her a thank-you message. Then I doze off again, while Louise whispers all kinds of positive words, to which she is accompanied by some jingle jangle floaty music.

Enlightenment

And I know that something essential deep inside me changed yesterday. I am so grateful that that foot fits right under my leg again, as it should be. I got lucky. Shit happens. Shift happens. It did happen. With a hard blow, but it's only a leg. No brain injury from a slap to my head. No spinal cord injury from a fall on my back. I made a huge shift, having fallen with my full weight on my ankle. It was not convenient, but how lucky was I? There are people, who first almost die or are completely dead for a while, before making a shift. We call that near-death experience, but when everything has been off, you are in fact dead for a while, so it should actually be called post-death experience.

The knowledge and experience that I could just fall at any time. I was used to that. I had been doing that for years. But the fact that I could just break something changed the urgency in my life. The ease of being able to rely on my strong body was gone for a while. Pretty arrogant to think that something like that wouldn't happen to me. And that wasn't even a conscious thought.

If you want to clear clutter from your past, you have to look for it. I always say: the quickest route to the solution is to go straight towards it! If you are afraid of something ... go towards it. But if you have long since lost track of what you are afraid of, then you can't get at it. And then you are helped by life or by something inside yourself. I slipped down a wet, freshly mopped marble stairwell. In a millisecond, my plans for that day and also for the rest of my entire life, changed.

My fall literally threw me into my deepest fear. The fear of not being able to go anywhere on my own. I had a sense of intense panic which I noticed in the moment itself. I was completely powerless, in optimal resistance and total denial. And all this went hand in hand with total surrender. There was nothing else for it. I knew I had to do something and I did something. I also felt I wanted to walk again and to carry on living. And I knew that I desperately needed myself to do this and that I was completely at the mercy of the people around me. Without others, I wouldn't go anywhere. A deeply moving realisation that caused fear, panic and utter peace. Knowing that I had no choice was somehow also very liberating. And reassuring. For a moment I didn't have to think about anything and at the same time I had to decide and do something now. HERE and NOW! An intense experience. A wonderful learning experience. You name it. It was all there at once and I knew it. On one level, I was having an intense experience and on another level I was talking the biggest nonsense. I told Elia that my foot had dislocated. You know? Like a dislocated shoulder. I knew I had really never heard of a dislocated foot. There was no such thing. My foot was very much broken but I was in denial.

Broken! A blind person was able to see that and Elia saw it too. But she let me ramble. She let me spout nonsense. She drove me to the best private clinic in Limassol and parked in front of the Ygia clinic. She hoisted me out of the car and sat me, majestically, in the sunlight on the marble steps. There I sat, wearing just one shoe and in my lap was that swelling thing that used to be my ankle. I smiled at a man walking past me and he asked if I needed help but Elia was already returning with a nurse and a wheelchair. Together, they hoisted me into it and wheeled me inside.

Elia stayed with me all day and signed as a guarantor for the bills. She let me ramble on until I became quiet and I had to admit that something was broken that I couldn't fix myself.

I will return to my trauma live show in the next chapter.

How about you?

Survivors tend to overlook issues. Especially things that hurt and that they believe they cannot resolve. It can take forever for them to admit they can't make it. But oh, how liberating it is.

16. Do you have desires you would like to give in to?

17. Have you perhaps found compensation so you don't have to give in?

18. What is stopping you from letting go and giving in?

Again, take some time with a pen and paper to retreat somewhere. And write it down for yourself.

BONUS 3

Deep down, you still have a big dream, but there's also a part of you that doesn't actually believe anymore that things can be different. Or that you are worth more than the life you are currently living. In that case, check out this bonus material.[1]

---

1) tapandletgo.com/bonus3

7. The search and the landing

"She felt a looming reluctance to ever land anywhere."

Dick Hillenius

Piet wrote down the above quote for me at the end of the intake interview. Piet was a psychoanalyst and taught psychology at the University of Amsterdam. After talking for an hour, he picked up an empty envelope, wrote that quote on it and gave it to me. I have kept it with me for years. I framed it and it's been on my wall for ages. I have sort of landed, yet when I look at that framed envelope I still think, "Yes. Still a looming reluctance."

Antonio (50) is a professional violinist and composer. He has travelled all over the world giving concerts, both solo and in orchestras and combos. He sent me an online request to help him, after he had watched all the videos on my YouTube channel. When we start the first session, he has just turned 50. He is completely stuck and barely able to go out the door. He is living back with his parents in North Carolina. And he is paralysed by anxiety and insomnia.

He doesn't dare go out of the house

He is trying to get off prozac and trazodone, but has not slept at all for a week. "What's your problem?" I ask. "Intense anxiety, stage fright, fear of failure, panic, insomnia," he says. "And fear of authority, like a conductor." And he is hypersensitive to sound, light, everything.

"I want to run away and don't dare go out the door anymore. I want to get rid of the pain." He is not suicidal. I immediately ask about that. Wanting to get rid of the pain is not the same as wanting to end it. And I want to know if he is a danger to himself. "I feel a deep sadness, crying about everything and worrying about my parents. A desperate feeling of loneliness, isolation. And I'm afraid that things won't work out. And that I will depend on antidepressants and sleeping pills for the rest of my life. Pills that don't even help."

During the first interviews, the story appears. Musical talent was discovered in Antonio at a young age. He is the eldest son of a family that fled from Nicaragua to the USA in the late 1970s. His parents did everything for his education at the conservatoire. All his time and energy went into playing the violin. He was not only afraid of failing and disappointing his parents, but he also carried the burden on his young shoulders of performing on all kinds of occasions. And to prove how 'free and happy' he was in the new country, where he did not feel at home at all.

A refugee child's fear cannot be ignored. Neither can grief and family traumas. Even if you are such a talented prodigy, which he himself had serious doubts about. Music had become his life and he played in renowned orchestras worldwide. After years of walking on eggshells, he has completely lost the joy of making music. He has not played his violin for months. He initially thought it was a midlife crisis, but it is only getting worse. He smoked a lot of pot for a while and it didn't really help. This triggered a kind of psychosis, making him panicky.

In eight online Skype sessions, we do a big cleanse. I teach him to tap on painful memories. And help him clear traumas and limiting beliefs. We replace them with other emotions and give new meaning to them. As a result, he starts to believe again, in himself, the world, life and his potential. Because he sleeps badly and wakes up early, we have sessions at 5am, in his time zone. That's at 11 o'clock my time. That's when the rest of his family is still asleep and he can tell his story uninterrupted.

Hey-ho! Out into the street!

And then comes the moment he has to get out the door. Exposure! Avoiding anxious situations is not a solution. On the contrary, fear is in your head, in your thoughts. And it grows in the absence of experience of a different reality. As I told you before: get out there! But then it is quite challenging for me that he is so far away. I can only talk to him at such a distance only through video calls. I can't go with him. I can't touch him. All those little gestures or an encouraging smile. Offering a glass of water, pouring a cup of tea. Handing a tissue or wiping someone's tears. These are the little signs of letting someone know they are not alone. That you are there. I do it automatically and enjoy doing it. Making a joke when I wipe someone's mascara. It comes naturally and I miss it during online sessions.

Nevertheless, my methodology works just as well online. Because first of all we take every step inside our head, in imagination, in fantasy. Because the body doesn't know the difference, so this way he can experience it completely. Every step he imagines he feels in his body, not just in the head. Yes, hypnosis works and you don't have to lie down or float around for it. It works if you sit up straight and are wide awake. We all do it, but mostly we don't know we are doing it. I use it consciously and explain what I'm doing.

First, I have Antonio describe in detail how to get outside. Starting from his front door, I let him experience the whole route. Whom he meets where as he walks down the stairs. I immediately intervene when this brings up emotions. I tell him to tap his face on every emotion. I give him precise instructions on which meridian points to tap. I keep the instructions simple and direct. I don't name meridians or muscles, but point to the spot. This encourages him to let go and has a relaxing effect. My way of talking makes him feel safe. He knows I am with him. This allows him to relive, live through and release each fear. In this way, I help him replace the fear with calming and empowering emotions. After all, I have asked him in detail beforehand what makes him happy. So I throw that in at the right time.

Just do it! The threat is inside your head

When someone says, "If I stay home the walls close in on me," I say, "Stay home tonight and wait to see what happens." After all, I know that walls don't move. Fear is inside your head. Confronting yourself with the situation is the best solution in such cases. You have to experience life and face your fear. This was also the case with Antonio. I guided him in 'trance' through the first steps out of the house. During this guided 'journey', Antonio met the cleaner in the stairwell. A tall man, who smiles kindly and encouragingly at him.

You don't meet that man if you stay indoors. Whether it happens for real or in 'fantasy' doesn't matter. It is an empowering experience. We are online in conversation during this walk, which takes place in his mind. And which he tells me about. I 'walk' with him, and images arise in my mind too. And I affirm everything that makes him feel good. And I help him tap away everything that is holding him back. This is how we change things on his emotional level.

"All struggles are fought first in the mind."

Joan of Arc

During one of his first trips outside his home, he finds a job. He becomes a driver of a children's train at a nearby shopping centre. At one time he would not have taken this job as he would have feared the crowds. But now his attention is focused on the children in the little train, who enjoy their rides. It does him good and it also makes him happy himself. He does not always manage to go, but he is getting there.

After the summer, Antonio takes some big steps. Literally and figuratively; a breakthrough. He picks up his violin again and even accepts an offer from the orchestra he played in previously. He is able to be a bit more light-hearted. Of course, I am thrilled with the result. I confronted him with his fears and made him swim in the mud. Sometimes that has to be done. Just long enough to want to stay out of it forever.

Sometimes his fears still grip him, but he knows he cannot run away from them. "Straight to it!" is my advice. He does that now, sometimes trembling with fear.

Regardless of where our fears originate, there is always a way to get rid of them. If you really want it and take responsibility for your own life. If you are brave enough to ask for help. Antonio wanted to be able to travel again and we worked on this in the sessions. He flew to Mexico, where he asked his girlfriend to marry him. A brave man, un hombre valiente!!!

Paid in music

Antonio was keen to give something extra and gave me the rights to use his music. He wanted to keep in touch, so we speak occasionally via Skype. He asked me to come to Mexico to give training and sessions. Fear of failure is common in his profession. And it weighs heavily on professionals, he says. And he wants such people to know how to find me.

In 2019, he returned to painting. "The violin was not my first choice, painting was my first love," he says. And he helps his wife with promoting her flower business. He told me his fears have completely disappeared and he has never felt better. "All thanks to you, Jeannette!"

In early 2021, he embraced his violin again with enthusiasm. And after I saw and responded to his new video series, he said, "Yes, I have a new surge of creative energy in my being!!!" He occasionally travels back to the States to play again with the orchestra, where he worked for years.

In his words

Between doing his sessions, Antonio wrote reviews on Google and Facebook, such as:

"Jeannette is a combination of knowledge and patience. And a genuine dedication to helping others heal. With her help and encouragement, I am on my way to getting my life back! She is a very skilled therapist. I was experiencing panic and fear of everything. But she was able to speak to me in a way that gave me the courage to take risks and face my fears. And when she saw that I was trying my best but couldn't do more, she was caring and empathetic. She gave me space to express my thoughts and even to disagree with her procedures. For she is confident and humble at the same time.

With Jeannette, I never felt like I was having a session with a therapist. One of those who follows a dogmatic series of standard models. She is fluid, flexible and intuitive. She feels the client's deepest needs and tunes in accordingly. I experienced her more as a good friend, to whom I effortlessly opened up with my deepest fears and doubts. That's another gift!"

The quest

We have to move on, so we build a survival system. And the 'best' way is to stop feeling. Not that it is really possible, but we are able to hide our memories. At least from our conscious daily brain. We move on to the order (or chaos) of the day. We no longer want to feel it. Because what good is feeling pain. It eats you up. It gnaws at you.

Without realising it, it gnaws away, even when we think we have forgotten it.

We adapt and we are given rules and punishment if we don't follow the rules. And there is a reward for those who do comply. From home, I was taught: always keep going, no matter what. And I did, as long as it worked.

If you are never alone, it is easier to deny underlying emotions. For some people, that's the 'solution'. Make sure you are never alone. Because when you are alone then you are with yourself. And you should avoid that at all costs. Or you are constantly looking to do 'nice things'. As long as there is no silence.

Drinking alcohol, often and a lot. With that, you also manage to deny underlying emotions better. You numb yourself in any way possible. And alcohol seems to dissolve entire pieces of your brain in the long run. Ideal, right? In our culture, this is accepted and even encouraged. Not officially, of course, but on the ground. Drinking alcohol is sort of 'normal'. And if you don't drink with us, you are not sociable. I've heard that so many times. "Come on now, have a drink and be sociable." And tea is not considered cosy at all in some circles. Then you stay far too sober.

There are many ways and means to deny your underlying emotions. I tried them all. I constantly distracted myself because I had to move on. I had to go on living, even though sometimes I really didn't feel like it.

Sometimes you yourself don't understand what is bothering you. Or that you have underlying emotions that make your life uncomfortable. Sometimes you don't know what to do with yourself. Or you just think a lot and know a lot. That can make you feel isolated.

I didn't realise what was going on. I had already done all sorts of things to clear my underlying emotions. I had been on a quest for years, learning to apply many therapies and methods. First to myself and then to others. And so I became a highly skilled therapist, trainer and retreat facilitator. I've seen it all, so everyone can tell his or her story. I don't scare easily, so with me you are safe with your deepest secrets. And I help you to just let them go.

Clearly, you also have to be ready to let go of underlying emotions. However hard you try, life cannot be forced. You have to be open to change and seize your opportunities when they arise. Various training and therapies helped me move forward. Sometimes they didn't. I also tried really weird things, because there is also a lot of nonsense on the market. With all the risks they involved. I did get a little better from almost everything. But the true turnaround came with a bang. BOOM! "She should just listen" said my three-year-old brother as the door fell shut and cut off the tip of my finger. That was over fifty years ago and he did have a point. My mother had already shouted several times for me to come upstairs. And I hadn't listened. The tip of my left little finger was dangling loosely back and forth.

The landing

(Continued of the Trauma live show)

I am an advanced certified professional in various modalities. I am a certified trainer and a professional hypnotherapist, a stress release and behavioural specialist. I don't often use these lengthy descriptions. But it is true. I can do it all and do it well too. I have studied trauma. I know how trauma is created, how to create and reinforce it. And how to get rid of it! I have been researching and studying how the brain works for many years. The mind, psychology, quantum physics, epigenetics, DNA healing, ho'oponopono and so much more. And now I am in the middle of this. And I am grabbing the opportunity of this experience with both hands. In the hope that it will remain a one-off particularly fascinating experience.

Elia and I met in Athens the year before. She attended the Level 1 training and I assisted with the seminar. Last week I was staying in her flat by the sea and I had already given her a few sessions. Actually, that day we were supposed to go hiking in the Troödos Mountains. She was already at the door when I fell down the stairs. So instead of going to the mountains, she took me to the clinic. She also tapped on me that day. Now she is about to give me a session. I asked her to do that because I want to release this trauma, as soon as I can. It's in the past and I don't want to repeat this over and over again. It's pointless, but I have flashbacks of the fall all the time. I tap on it when it happens.

The worst is the image of the foot. When I see that image in my mind, I can feel the pain again and a shiver goes through my entire body. Every time. And that image keeps coming back, without me choosing to think about it. There is no reason to keep that fear and shock within me, so let's work on that. Elia wanted to do the session with me on Friday, as she has more time then, but I asked her if it can please be done today. And no, I absolutely don't feel like it. I never feel like having sessions myself, but I know it works and it always gives relief.

With everything I can, I also know that I cannot do this alone and I am so grateful that my hostess and friend wants to help me. I trust the process completely and I know she knows what to do. It is a pretty clear and simple process. I could even write out the questions and ask someone in the street to ask them to me. There would be the results in the street as well, but it still feels safer with someone who is familiar with and dedicated to this work.

It's important to be guided by someone who is not afraid and who stays with you no matter what. It's even more important than that person knowing all the techniques. And the more you understand about how the process works and the more experience you have, the better it works. I also know that the more often I used it, the faster I became at guiding the process and understanding it better. Because it's not necessarily important for the counsellor to know exactly what happened, nor understand exactly what is going on in the client's mind, which is a misunderstanding among many therapists.

The subconscious is much faster, so you can just say to the client "Go there". Then you can see in the face, body language or breathing whether something is happening. And through the words they use, you are able to hear when they are there, in that memory. And I don't need to know exactly where they are, I just check how strong the emotional reaction is. Then I guide them in letting go until it's totally gone. That is the process, which I love so much and have become so good at. Today it's my turn and I feel like it, while I also don't feel like it.

Finally letting go

Because I know the process really works when the client wants to change, I just trust it. I've used it so many times that I've lost count of how many sessions I've given. I stopped counting. I know Elia knows enough to give me a session. If she wants to use other methods too, that's fine with me. I know she is familiar with Louise Hay and has done other training. So it can only get better and I need help. I tell Elia that I trust her completely and that I know she is good and there is no reason to delay the session.

When I hop into her practice room, she has 2 chairs ready: one for me and one for my leg. There is water and we are ready to go. I ask her, whatever she is going to do, to go to the incident with the foot. I ask her not to waste time on an extensive intake and all the hassle around it. "Okay" she says. And she asks me, "Why are you here?" And I say, "Because of this." And I point to the foot in plaster in the chair. I want to change this trauma, this memory.

She asks some questions, which are good, but she doesn't give me time to answer them. So I say, "You're doing good, great questions, but please go slower, so I can find the answers." I know that the fact that I am slow is not so much because of the language, but because I am kinesthetic. So I need time to describe in words what I am feeling. She asks something and I feel I am my father, who sees his own father as a gigantic force. Very calm and very peaceful. My conscious mind thinks, "Huh? That was never the story?" But it feels right, so I leave the thought of the story for what it was. Elia quickly asks questions and I say, "I don't know, but this looks good and feels good, calm, very strong. There is no problem here."

In response to another question, I tell her about falling a few days ago. And about how I sat on the stairs and looked at my leg and thought about what I said to her over the phone. "I am on the stairs and something is completely wrong here." And then she asks me: "When did that happen before? Do you have any memory of that?" She asks if something like that has happened before and if I have a memory of it. And in less than a second there is the answer. I feel huge resistance and I think, "No not that! I do not want to go there!", but I say, "Yes." Because I have exactly such a memory: being on the stairs and knowing that something is terribly wrong.

Sobbing heavily, I tell Elia what I remember: I was just 14 years old. It was a Sunday evening. My sister's boyfriend had just left and the doorbell rang. I thought he had come back, until I heard my mother screaming. "Jeannette, it's your father. Call the police." It was my father, accompanied by two of his friends. Two guys who watched as my father tried to pull my mother out of the house and throw her down the stairs. At least, that's how I always understood it. That was his intention. He came to destroy my mother.

It was utter mayhem. My sister and little brother heard the noise and came running. I screamed and called the police I don't remember how many times. And it took them forever to arrive on the scene. My mother could have been at the bottom of the steps a hundred times, but she fought like a lioness. She pushed him off her and got the door closed again. But he kept pounding it and came right through the door. Not quite, as we were all hitting him. My sister had the presence of mind to take the bread knife out of my hands. I'd fetched it from the kitchen, she later told me. But sticks, tubes and a granite ashtray were all considered acceptable.

We fetched all manner of weapons to arm my mother. And we really let rip as well. The granite ashtray, a Father's Day present, delivered the final blow. And before he collapsed, my mother pushed him back through the front door that had split apart. The police picked him up and took him to the hospital. I would never forget those cowards who stood by and did nothing to help us. We were children of 16, 14 and 9 years old. In fits and starts, I spat out the memories. Whilst Elia tapped and tapped, "Let it go. It's time to let it go. It's safe to let it go." I don't remember how and what she said. But I hadn't cried like that in years. It came from very deep and it was immensely liberating.

Respect for emotions

What was so liberating, in the end, was acknowledging that, in spite of everything, I loved my father very much. And that I had never realised it before. I spent my life hearing how awful he was, what was wrong with him and how much he and I were alike. I spent years listening to my mother, who needed someone to tell her story to. My father was wacky. He would be the last one to deny that too. Emotionally, he may never have moved beyond being a five-year-old neglected toddler. Because that's what he once was. And that experience determined the way he dealt with himself and the world. That is my interpretation, because I never asked him. And when I asked him questions, much later in our lives, he didn't quite get it. He tried his best to answer honestly. But he had never thought about the past or his feelings in a way I could relate to.

After the fight, my mother let another troublemaker into our home. Because she was afraid my father would come round again. Understandable, but she brought an even bigger problem into the house. There was no place or space for processing our fear and pain. We learnt to laugh it off, sang silly songs about it and carried on. Like we always did. This experience of violence was not the first one, it all started when I was much younger. Nobody asked about it and nobody talked about it anymore. The door was fixed and we had no idea that we might need fixing. At least, we had never heard of that. At twenty, I opted for social studies. I received intervision and supervision and studied human beings and human behaviour. I started working in psychiatry, crisis care and child protection, which kept me stuck in the same shit for years, but at least I learnt that you could and should talk about it.

This is how I tried to come to terms with myself and my traumatic memories. At the same time, I was surviving by trying to forget. But that never worked. If seven streets away someone raised their voice, I went straight into fight or flight. Even though I knew it wasn't about me. If I thought someone somewhere was being threatened, I always felt I had to do something about it. That's why I worked in crisis care. Not really a choice, but rather an automatic response. I was doing what I was good at, but it wasn't any fun. I didn't know that. I was just my 'normal' job.

Always threats and always having to help. I had special antennae for that purpose. When I was sick and tired of it, I decided to quit my job. I went to Greek classes for a few months, bought a ticket and left for an indefinite period of time. One of the best decisions of my life.

Are there any prerequisites for hypnosis?

Rob de Groof:

You can only solve something with hypnosis if the person himself is ready for the change. If they are not ready for the change then nothing is going to happen. I want someone to come to me and already know what change they want. If there is one thing I don't like doing is working with consciousness. I want to hypnotise people and work with the subconscious. I also don't do intake interviews or pre-talks. Then you can't address every problem because the client has to have a very clear view of what the problem is. And if we first have to go into 4 sessions to find out what that actual problem is and that they may or may not want to change then they have to go to another therapist first.

I recently had a client who I had done two sessions with me, around alcohol abuse and he sent me a message that he needs help again because his wife caught him drinking. And I replied to him that if he wants to dry out completely, he should discuss that with his doctor, because drying out completely is not so good for the body. And I say, "If you just want to stop drinking because your marriage is under pressure, then there is nothing I can do. If you are caught drinking then you are not ready to stop." You can do a lot with hypnosis but the person himself has to be ready for the change. I think that is the most important thing.

Ina Oostrom:

In someone with multiple personality disorder or psychosis, there is a risk that symptoms could get worse. If people are on medication, I always tell them to contact their GP. If necessary, involve the treating psychiatrist. More and more GPs are familiar with my therapy. There are also more and more referrals for physical issues, such as irritable bowel syndrome and countless other complaints. Those are mentioned in the psychology textbooks as being suitable for hypnotherapy. I deliver training to doctors, nurses, midwives and psychologists.

Kim Jewell:

I always inform my client that they're in charge of the session. That at any given time, what you and I both know, if we want to help the client change something and a part of their psyche isn't willing to accept that, the mind will not do that and that's where it becomes really safe. But what I like is the way that we do that: what we do is giving the client the empowerment to do that for themselves. So therefore they also know they're in charge of the session, they can stop it at any time, they don't have to go and re-traumatise themselves.

In fact what we're actually doing becomes very empowering. They come out of the session, saying: "I feel lighter, I feel stronger, I don't even know why that bothered me in the first place." And I always check before we go there what they want to replace it with and how will they know that they're able to do that. Because quite often the survival mechanism, that tries to keep the resistance in place, is there, because they can't even imagine life without it, so we have to help paint that picture of how it can be different.

Martijn Groenendal:

Yes. If someone, as soon as they talk about their trauma, has a panic attack then you have to put some intermediate steps in place. Firstly, make that person feel more powerful and then make them look at an old situation. Think about dissociation, about safety. You can ask, "What do you need to be able to work on this? What will it bring you when you no longer have this problem?" So, first work on the mindset about the trauma. That sometimes needs to be changed first. Also the mindset gained in the trauma. In addition, you should be able to reassure someone. That way, we can look at what happened in the trauma.

How about you?

Giving your trauma a place and learning to live with it is crazy. You don't say to someone with a broken bone: "Give it a place and live with it." No, you clean it, you fix it, you heal it." The same for a broken heart. Learn to revive, review and revalue the story. That is not difficult at all, because those negative emotions pop up by themselves all the time.

That's why we have to drink so much when we don't want to feel it. The emotions are there for the taking. And when you have found them you can let them go. + PLUS (important!) immediately replace them with different emotions. Like laugh about it and joy and so on. Seriously. You can do it. Doesn't bother anyone. And you get better. Literally much better.

19. What emotions keep popping up in your life or in your mind?

20. Do you find yourself unable to push them away, no matter how much you drink, snack or run?

21. Imagine what your life would be like if you could cut out these emotions.

Sit down for a moment. Write them down or paint or draw them out.

8. Solution

Somewhere, at some point, my sister and I agreed that we would dive into this life together. And that we would then stay together forever. Noone would ever think that we were sisters. Except for our parents, because they were there when we arrived. And grandma too. And my sister also knew when I finally arrived. She'd had to wait 2½ years for that to happen. Not that we have any proof of the original pact between us. And we don't remember it either, but we think it's a good story. And I'm inclined to believe it.

We are polar opposites and are nothing alike. People don't see it, but they hear we are sisters. Our voices sound almost the same, people say. We have shared a lot. When she had cold feet, she would crawl into bed with me. And then when she was sleeping, it became too hot or cramped for me. And then I would get into her cold bed. When my sister decided on hairdressing school, I was screwed of course. Imagine how useful it is when you then have a sister with real hair on a real head. One who is always up for an experiment. Afterwards, there was often crying and arguing, but that never lasted long. My hair has been all colours of the rainbow. And one day, I came into school with so many curls that no one recognised me. We looked after each other's children a lot and sometimes they didn't even know where they were exactly. We did all sorts of things together. Including going on holiday together with our families. And we will probably do that again.

When I learnt to tap, I wanted to practise my newly acquired skills. I tapped on just about anyone who was within reach. I wanted to test if it was really that good. And it appeared to be. So my immediate family and friends were the first to have fun with it. I found my mother crying after my first seminar in Budapest. Her little dog had died and she couldn't stop crying. All day, all she thought about was the dog's last look before it was given an injection. And she felt guilty that she had her put down. But the little dog was old and spent. After the session, she thought only of the wonderful memories she had with the little animal. All of which I had made her tell while tapping.

I never know exactly what results I achieve with people. Sometimes they tell me. But even I don't know what exactly changed in my life, nor how it happened. And that doesn't matter either. All that really matters is that you know and believe that you can change. That you can make the decision to do so and then just go and do it. And then it always works out. It's a matter of persisting long enough. Work on it often. Never give up. Keep going. Just let go. When the grandchildren told my mother of a problem, they were sometimes referred to me. My daughter sends her friends to see me. And, of course, I gave sessions to my sister too.

Been to grandma, grandma fun!

Some things take a long time and sometimes the timing is right to flip the switch all at once. My sister's weight had fluctuated for many years and she was often overweight. When she held her grandson in her arms, something fundamental changed in her. She thought, "I have to be able to keep up with this little boy and I can't do that with the body I have now." She stopped drinking wine, started eating better as well as exercising with a personal trainer. She lost 30 kilos in no time. Her grandson spends at least one day a week with her. And regular sleepovers are held at grandma and grandpa's. That cute little guy keeps up quite a pace, but grandma always goes along with him. And she's able to keep up with him. "Been to grandma, grandma fun" he told me recently.

Apart from being 30 kilos lighter, my sister is my real sister again, the way she used to be. But an even better version. I can't really explain it, but I tell her every time I see her. That I am so proud of her. I know the journey she's been on, even though, of course, we both have our own story. And our own perceptions. That also means it is never boring with my sister and we never run out of things to talk about. I tell this story because I realise, especially with my sister, how much we can change. I know her pain, for as much as you can really know someone's pain. And I know what she was like as a child. And over the past 50 years or so, we have walked side by side. And when you empathise with each other and see each other regularly, you see her as she is. Or as she's becoming. It's so gradual that you don't really see what could actually still be possible. What I am trying to say is that it is still a kind of miracle now that I see my sister without those 30 kilos. And she notices it herself. It's not just the larger clothes or the extra kilos. It is as if she has let go of a big burden. Her voice, her eyes, her skin, her face, her talking, her laughing. They 're all fresh again. Brighter. And she herself is increasingly clear in what she does or does not want. And is able to do. And what is still possible.

That's how fast things can change, when you decide enough is enough. All kinds of ailments have disappeared like snow under the sun. Too bad we get used to things like that so easily. And too bad we can get a pill for everything from the GP. Someone in that position should take the time to ask some sensible questions. For instance, about how much you drink. I ask all my clients. It's not a difficult question. Unless you don't take the time or you're not interested in the answer. But you could ask if someone who drinks might want to change that. And that there is help available. That you don't have to do it alone. The last time I saw the GP it was about whether my cholesterol was too high. And he asked if I ate a lot of fatty meat. No I don't. And then he asked whether, being a vegetarian, I miss the sausages from Hema. Seriously. True story. He is now retired. I got a prescription for cholesterol reducers, read the leaflet and returned the meds to the pharmacy. Never went back.

I see people with walkers and mobility scooters, which I don't think they should need at all. In fact, I have a neighbour who has one and takes it for a ride about twice a year. The rest of the time, she takes the bike or goes along in her neighbour's car. She's still able to walk very well. Look, my 84-year-old mother didn't want a mobility scooter at all. But after being run over, she was less able to walk. That's a different story. She still feels bad that she can no longer cycle and misses the freedom and independence. You know, the daughter of the man who always said, "Head up, back straight and keep going." May he rest in peace. Or he has long since returned to this planet. No time 'to rest in peace'. There is still so much to do on this earth.

Back to my brave sister for a moment. Being able to observe it so closely allows me to see what happens when you change. Of course, I see it in my daily work, but I haven't known those people for long and haven't been close to them. Over the past 25 years, I have seen my sister gain a lot of weight and I have often worried about it. But everyone is on their own journey and I have learnt to keep my mouth shut more often. Although that doesn't always work with her. But now I see it happening before my own eyes. It is true. There is no need to be a physical wreck when you are 60. You don't have to smoke and drink and snack, as was normal at our house. You can also face your problems and solve them. You can choose to take care of yourself and your health. You can learn to start loving yourself. Get clarity. I don't just see it. My sister also tells me. And I'm so happy about that.

How did it come to this with my father?

I can only guess of course. He has been dead for years. And I wonder if he ever understood it himself. That handsome man whom the neighbours liked so much. When the fathers of girlfriends fell ill, I thought, 'Why him?' The child would then be scared of losing her father. And sometimes that's what happened. Then I thought, 'Why does mine stay alive?'

During the time when I still thought I knew better, I analysed him a lot. And often thought and talked about my father. But, of course, I could never answer the question at all: How did it come to this with my father? Nobody knows. The best I can do is live with a good story about my father. And a good story with my father. Not that I'm going to deny what happened or what used to cause pain. But I can let go of the emotion, so it doesn't bother me anymore. I was able to forgive him and strangely enough, I started forgetting miserable episodes as a result of that. My life has become lighter. Not because of my father, but because of what I did myself. No one else is able to do that for me. It is my job and my responsibility to make me happy.

Of course, I too once wanted a father who would make me happy. Or who had at least tried to. But my father didn't have that in him. He didn't have it. And I don't know exactly why either. I can keep whining about it, but then he still doesn't have it. And he's also very dead now. And I think that what he missed or lacked was absent before I was even born. So clearly, it had nothing to do with me and I'm able to accept that. It's a lot easier and I get immediate results. I've been healthier ever since, I sleep better, I breathe deeper. I like my nicer and kinder self a lot more. Discovering that I loved and missed my father was liberating. One of the best experiences of my life. Crying deeply and letting go of the sadness which I didn't even know was there.

Even if my father was still here, I don't know if he would have a meaningful answer. He would try, though. When I was about 30, I was with him one Saturday night, having supper at my sister's . I had just come from an intensive psychology training weekend, where there was always a lot of soul searching, crying, laughing and dancing. My first thought was, 'Jesus, I really don't feel like doing this.' But immediately afterwards came the next thought: 'This is an opportunity.'

144

He'd had a drink and was quite talkative. And he was happy to see me. And less wary. I was not the only one here who was a bit suspicious. This little apple did not fall far from the tree. And here sat the tree, slightly inebriated. He was perhaps even more afraid of me than I was of him. I had not exactly shown myself to be a fan of his. And he had done very little to make me a fan. My father was a very insecure man. He had started drinking for a reason. Drinking had given him an outlet. Not convenient, but it worked. I believe that, emotionally, he was a mess. But who says I could even know that? Maybe he felt he was doing ok by his standards?

The interesting thing about that night was that I sat down with him at the table. And I started asking questions. And to my surprise, he gave unexpected answers. With slurred speech, but better than nothing. When he left, he'd probably sobered up a little. My sister served him cups of coffee. Maybe too many, but so what? He actually started a conversation himself by asking about my work. I worked in Child Protection. And he asked if I had to go after types like himself. Fathers who wouldn't pay child support. I explained to him that the finance department did this. And I didn't work there. I explained that my job was to advise the Family Court. And that I got involved with dysfunctional families. Where children had to leave home because of violence. And that I had to advise on access arrangements and custody in divorce cases.

Blatantly honest

He seemed to find it interesting and asked all sorts of questions. My father was crazy, but he wasn't a good liar. At least, not to us. He

145

might well have lied to himself about everything and, like his father, he was a bully. I didn't understand him either. And felt a great distance between us as a child. But when he'd had a drink, he was completely honest. So honest that he forgot he was talking to his children. My sister and I would sometimes look at each other and raise our eyebrows. Surely, you don't tell that to your daughters? Any sense of shame left him when he was drunk. That was awful. I was mortified when we had to try to get him home when he was stone drunk. He wanted to walk into people's houses. Or he urinated in the street. Or worse, in the front gardens of our schoolmates' houses.

But on this Saturday night, I found his honesty very reassuring. Because, prompted by my job, we talked about domestic violence. And its effects on children. Not that he was concerned with that. He told me that I had written him a letter once, saying he had done everything wrong. I had long since forgotten about that letter. And he told me a lot more. And his perception was that it was all me. I was angry with him. And he still didn't understand why.

After all, my father was a victim. He just plodded on and I think he felt sorry for himself. He told me that at one time, I was walking down the street when he cycled past and I had ignored him. That was shortly after he had broken down our front door in a drunken state. And he was indignant that I, a child of 14, had not waved my hand at Daddy.

It never occurred to him that I might be scared. And that I had no idea what to do either. I felt terrible that I hadn't said anything and hadn't waved to him. I was just startled when he suddenly cycled down our street. But he had told his mother and she in turn had complained about it when she ran into my mother. My father seemed unable to see a connection to his own actions. So what can you expect ? And I had somehow always known that.

I could write a whole book about my father. It would be an interesting book and good fun. Because even now I sometimes burst out laughing when I think back to all those incidents. He had a speech impediment (he couldn't pronounce the R sound), so we made fun of him. When we dared to, so not when he was around. It's a little sad, when your own children make fun of you. But yes, that's how we were brought up. If you call that an upbringing. Not an excuse, but a consequence.

I made peace with my father. Before he died, I still annoyed him sometimes. But at the end of his life, there was peace and afterwards, it embedded itself in me. It's all about perception. And now, I have a better story about my father. Fortunately, because I often hear that I look like him. And so it's better to have a good story.

My own images

On a sunny Friday afternoon, it took just an hour for my father's house to burn down. How the fire started, we will never know. We saw the destruction and the utter mess it caused. Everything black and a terrible stench. My sister and I filled a few bin bags with the most embarrassing items of rubbish. We discreetly dumped them in the dustbin. I will spare you the details. His fingerprints were melted into the armrests of his wheelchair. He was in hell even before he died. So hopefully he was allowed to skip going there after his death. The only thing in his house of value was his quality tools but these had been reduced to a molten lump of metal.

My sister and I were waiting in the corridor of the Burns Unit in Beverwijk, before we were allowed inside. My father was just being changed. The corridor smelt like an eel smokehouse without eels. We stood side by side opposite his room with our backs against the wall. I asked my sister, "Do you actually know what he wants? Burial or cremation?" My sister's reply: "Well, I don't actually know, but cremation I imagine." We almost choked on our laughter which was hardly appropriate of course. What we saw there was not great, but I never quite forgot the smell, until I started tapping on that too. At the time, the nursing staff advised us to take pictures of him. So we did. But I see no reason to look at them anymore. I have my own images to remember him by. But not of his last five weeks when he was wasting away. In my hallway hangs a photo of him as a boy of about 14 years old, next to my own school photo. Two peas in a pod.

After about three weeks, my sister, who knew I was going to Beverwijk, called me. "Don't be shocked, but daddy has his eyes open again." That was because he got scabs on his eyelids, which caused them to pop open. Sadly, he couldn't close them himself. According to the doctor, he was not able to see anything and was not aware of anything. So I started swaying my upper body from side to side right in front of him. His eyes followed me from left to right and back. And tears ran out. Just a coincidence, according to the doctor, but still. After four weeks, we pleaded with doctors. This hell had to stop. Time to go to heaven, which he did not believe in. They stopped giving him the astronaut feeding, but he lasted another week. Tough rascal.

Whilst on his deathbed, my father gave me five weeks to do whatever was still needed to be done. For example, to look at him very carefully. And then I looked under the sheets. Yes, that was suddenly possible and I was curious. I don't pass up opportunities when they arise unexpectedly. Just like my father: shameless. I saw how good his skin actually was. How young looking! Well, the healthy, unburnt skin at least. And I had the time and opportunity to hold him. And to put my hand on his heart. It was beating. I took my music score along and sat by his bed practising. So I sat next to him and in front of him singing. All heartfelt religious music. I was able to say anything I wanted and I also wanted to tell him about all sorts.

Before, I would have told him to drop dead, but now I said he was allowed to die. And that I loved him. And that sometimes, I even seemed to understand him. And I thanked him for giving me life. With strong genes, a strong body and a good brain. And a sense of humour, because you need that when you are his child. He had that himself as well. We didn't exactly share the same jokes, but a sense of humour to help cope with life. And I told him not to be afraid. It couldn't get any worse than what was left in this life. His body was broken. And even if they tried to fix it, he would never be able to be or do what he wanted again. We pleaded with the hospital not to patch him up any more. He had adjusted quite well to living in a wheelchair after suffering minor strokes and other ailments. He did not want to continue living like a vegetable with only two fingers left on his left hand. And the right side was already lame. He had often been lame before, but that was over by the next day. This paralysis would never go away.

Dad, the priest and the police

I was not entirely surprised when I heard about the fire. As long as my father was alive, there was unrest in the air. And once my father was dead, I felt relieved. The unrest that clung to him had never left me. Even though he had been semi-paralysed for years and I knew he wasn't able to do anything anymore. That restless feeling, not knowing what was hanging over your head, had never quite gone away. Until I saw he was dead. Until that day, I took him food every week. Then he'd say "Hello my little Jeannette" and I would give him a kiss. And then for a moment, it felt like old times when life was normal. Only, it never was, but I didn't know that at the time. I always put garlic in the food because I like it and it's healthy. He didn't like garlic and once gave me a whole salami. "You take that, because it has garlic in it and I don't eat that." He had his ideas, but as long as you mixed his food up thoroughly, he ate everything. "Just eat what you're given" I thought and then he would sit and happily eat the lot. He always said it was delicious. Isn't that sweet?

The Burns Unit in Beverwijk is a Catholic hospital. My sister had received a call telling her my father had died and so we went over there on a sunny Sunday morning. It was four days after his last birthday. Without cake, beer or peanuts. He looked rather weird, which we'd been warned about. But we were used to that by now. We were asked if it was OK for the priest to stop by. "Sure," my sister said to the nurse. And to me, with a broad smile: "There's no harm in that, do you think so?" We had a good conversation with the parish priest about how and where my father could be buried. "Just throw me in the sea" he had said when we had talked about it at some point, "Or just feed me to the dogs." I told the priest about that and we all agreed that his last wish could not be fulfilled. It's illegal and not very nice towards the dogs, the sea, the planet, us. There was another man in uniform in the room that morning. He turned out to be a detective, because fire is an unnatural death and requires an investigation. We were not suspects. It was a formality. But all in all, a weird last meeting with my father. We were sitting next to his bed, chatting to the priest. Dad, the priest and the police in one room. Bizarre.

You are the solution

My solution did not come from my father, but from me. It came from my desire and willingness to make peace with my father and with myself. The solution is always within yourself. Whether you want it or not, there it is. "Be the change" said the Mahatma. It was and still is a quest, probably for as long as I live.

My life has become so much better after I rewrote the story about and with my father. The story keeps getting better and more beautiful. Remember the example of heating up the chicken soup? I released and let go of my pain and since then all sorts of beautiful memories have surfaced. That's all I can do. He was buried in Beverwijk, where he died, about an hour from where we lived. We were not at his funeral, but that's a story in itself. There is so much more to say, but not now. PEACE.

Can anyone do self-hypnosis?

Kim Jewell:

Absolutely we can! And this is one of the reasons, I think, we both like using the self-hypnosis and the tapping that we use as a tool. So that our clients can help themselves start to bring off that fight-flight-freeze-charge and allow them to. When I explain to my clients what's happening when they're being triggered by trauma, I explain that in the analytical mind the blood flow is drained and so they're not thinking through their conscious mind anymore. They're now tapped into the emotional state of the earlier event, whatever that was. When we teach them how to use these tools that we have, they then can reduce that sense of fight-fight-freeze, allow the blood flow to return to their conscious mind. And so therefore they're allowed to be in the present again and they feel more empowered to deal with whatever showed up that triggered them.

Martijn Groenendal:

As soon as you're breathing, you can learn self-hypnosis. Self-hypnosis is when I apply a process to myself, usually with my eyes closed, giving my autopilot commands, suggestions and ideas that work to my advantage. We have all kinds of techniques for stress resolution, for weight loss and for other specific applications. Every thought is an auto-suggestion, sometimes feeding a pattern that I don't really even want. And which most people are not aware of. They think, "It's just a quote"; but a suggestion from myself to myself is a self-hypnotic phenomenon. People don't associate hypnosis with thinking. But once we think a thought, which touches our self, we are touched in our heart, in our subconscious. That is self-hypnosis, whether it works to your advantage or disadvantage. Of course, we will assume for a moment that all people who engage in hypnosis do so for the positive. Then the landscape of self-hypnosis is: always learning to think in a way that works in your favour.

And taking responsibility for your thoughts. Monitoring your own thoughts, even in silence. Then you are disciplining yourself, which is not such a popular concept. In this sense, self-hypnosis is a way of disciplining our thoughts, feelings and behaviour. By definition, we then have to go off autopilot to choose something new. Ideally, you need a mentor or proper training. If someone has little knowledge and little experience of it, it may be too abstract.

I once had to choose a joyful memory at an NLP training. And then I had to discover whether that image was in colour or black and white. The idea that we record our own thoughts in a certain way was already a huge revolution for me at the time. For good reinforcement of self-hypnosis, people need to be trained by you or by me.

Ina Oostrom:

Yes, self-hypnosis to relax, but if you want to explore within yourself, you need your analytical thinking. You can't take yourself in and out of hypnosis. For that you need someone else. We started a community, where people can find all kinds of information online for free, including self-hypnosis audio files.

Rob de Groof:

In symbology, we 'install' something that they can immediately do themselves. That is a form of self-hypnosis, with which they can very simply enhance that feeling. I think it's recommended for everyone to learn hypnosis themselves. It is a tool that you can use very easily and very effectively. Even if it is just to make all your days feel good.

How about you?

When you dare to let go of old stories, there is room for something new. Funnily enough, if you have a little patience, the new story usually arises naturally. It happens in every session. As we tap out the pain and let go, the picture changes by itself. You just wait for it.

22. What new experience is already waiting inside you?

23. You've probably experienced laughing in a crying fit. Or vice versa.

24. Our emotions are close together and replace each other with ease.

Sit back down with your pen and paper in a good place where you won't be bothered. And write down the answers. And also anything that comes up in response to these questions.

BONUS 4

Download the prerequisites for successful sessions with quick results and lasting change.[1]

---

1) tapandletgo.com/bonus4

9. As long as my heart keeps beating

Judgments are a hardening of underlying emotions which we deny. We live in a culture, where it is common to compete, fight, laugh at each other and humiliate each other. This is what we are taught. Small children don't do this of their own accord and a lot of people don't like doing it. Bullies are the scariest kids in the class. The less you feel loved the more you have to prove yourself. As I wrote earlier in this book, we now know that excessive criticism of others and constant self-criticism are symptoms of trauma. If you do this yourself, you know what it is related to. And you need to seek help to resolve the issue. It won't go away on its own. That's where this book comes in.

Maybe everyone has an experience of being bullied or suffered from some weirdo's hangups. So have I. Sometimes I could feel how the bully was in a bad place and that made me not hit back out of pity. I didn't want to make it worse and I wanted it to be over as soon as possible. Just like my mother did. But sometimes things went too far for me and I did have to react by doing something nasty in return. I think I've always been quicker with my head than with my hands, so my words are my strongest weapon. I can bring someone down with two sentences. And I have done that. But I no longer feel the need for it and have learnt that doing nothing and saying nothing is often better.

Then I don't have to regret afterwards what I did or said. If it's not really necessary, I don't do it anymore. Sometimes it felt necessary.

Small children don't do it naturally, so when children start bullying, it is, I think, a result of their own frustrations. I don't remember my own child ever bullying anyone of her own accord. And that has to do with her feeling loved.

The toughest project

My stepfather had a bigger influence on me than I wanted to admit. And actually, I didn't even want to mention him in this book, but it's a good and useful story. Because of him I noticed at an early age how mentally strong I was. Bullies are often the scariest kids in the class. So was he. A scared shithead in an oversized body. A sneak. My crazy family, I couldn't avoid them and was part of them. Not with this one. He was eager to let people know how powerful he was, but I pretended he didn't exist. I was 15 and I was very good at keeping quiet and ignoring. I looked past him, I talked past him and I lived past him. He couldn't stand it, but I kept it up for months. Much later, I heard that he couldn't keep his hands off children and I had known that, without knowing what I knew. The distance I kept protected me, because he never dared to touch me. I don't know what I would have done, but he was more afraid of me than I was of him.

Let him rot

He has been dead for years, dying of an unexplained illness, sort of slowly rotting away, which I found reassuring. Everyone gets what they deserve, I thought. I dreamt once that I killed him with a potato peeler. I put it straight into his chest and suddenly wondered if such a small knife would do the job. But he stopped moving and died by my badly armed but powerful hand. It was quite a job and not at all easy, but it felt very good. I think I was about 17 at the time, but I have never forgotten that dream. It was a real-life experience. Shortly after the dream, I packed my bags (actually I grabbed some big bin bags) and left. I never saw my stepfather again. A few years later, my mother also left him.

I have long since ceased to be that child and the past is over. When I came to understand that hate does me no good, I started to let go of those memories and emotions and forgave him. He hadn't made himself either and had had a rough past. That's no excuse, of course, but it made me realise that it had nothing to do with me. People can act fairly moronic towards you, but that doesn't mean you deserve it. He was my hardest forgiveness project. I didn't do it for him, I did it for me. I just needed to let it go. In The Hague, my town, we say, "Boy, just let it rot."

There was more going on at home and in the neighbourhood, which meant children were not safe. Not that they were all bad people. There was some real scum among them in our street. But most of them had no idea who they were themselves, and as such had no idea who these children were or what they needed. And that lack of safety was compounded by abundant amounts of booze, making them all 'weird'. So I had to make sure I kept my distance, even if I didn't understand why.

So to survive this and to protect myself, I was not at all nice, not friendly. I didn't laugh at the jokes and avoided as much as I could. I listened to no one but my inner knowing, to that deep silent voice.

"A man who knows more becomes lonely."

Carl Jung

This actually causes a sense of loneliness. I learnt to live with that from a very young age. There was a great distance between me and the world around me, because I didn't feel that anyone understood what I knew for sure. And I couldn't tell anyone, because I didn't know what I knew. It was just a feeling. It meant not being liked. So from a very young age, I couldn't care less about what others thought of me. That, I think, is what my mother called being 'different'. And that's exactly where my answers are. In that maladjusted child. I was sweet, but not very friendly. I was not easy going, preferring to withdraw. I was sometimes called a loner.

Kim Jewell said, "Emotion is our first language. Children under five are so present and honest and they are so real. And we lose that when we start trying to perform. Once we have that analytical mind and try to get people's approval. We try to figure out the best way to behave."

I just had too many adults around me in my younger years, whereby I didn't feel at ease or at home. It wasn't a conscious choice. It was mostly a feeling, but it was too strong to deal with. It bothered me, but I couldn't help but navigate based on that feeling. And 'growing up' never quite worked out in that respect. Nor was my blunt manner ever a choice. I tended not to realise this until it was too late.

But what seemed to be my long standing problem turned out to be a useful quality after all. There were school teachers, family members and others, who found my boorish directness quite irritating. Nowadays people, especially my clients, consider it to be my best attribute, helping them put things in perspective and letting go.

Familiar is not the same as safe

I found safety within myself, travelling on my own in Greece. As I slept under the open sky on a remote island, I suddenly experienced how wonderful it was to know that there were no people. It was deafeningly quiet. Occasionally, I could hear goats scurrying around and here and there something rustled in bushes nearby. I knew there were snakes and scorpions crawling around the tree I was lying beneath. That tree stood on a beach at the tip of the island. I had walked for hours with a backpack, which contained mostly large bottles of water. Because there absolutely nothing happened. All evening I sat by the fire that I'd lit. That had now died, except for a few glowing embers. The light came from the stars, lots of stars. I looked across the dark sea and saw a fishing boat with a single light far away. There I lay, naked on my sleeping bag, under a thin cover.

And I realised I had never felt so peaceful and safe. Far away from civilization, far away from people.

Even though I hadn't really known why I was doing it, I had prepared for this trip and I had quit my job so I could go. Most of all I knew what I no longer wanted and I knew where I wanted to be. I wanted to be in Santorini and beyond that, I had no idea. I went with the flow and set off. Impulsive and ill-prepared as usual. Nevertheless my solo adventures did me a lot of good. I spent hours, days, weeks, months staring at the sea. I spent a week on Santorini, wandered the islands for months and ended with five weeks on Gavdos. A rock in the sea where there was nothing to do. Not then. Thirty years ago. I still stare out to sea. That never went away. Been doing it all my life, over and over again. Breathing deeply in and out. Enjoying letting go.

Freedom is found in new experiences. In walking new paths. Doing something you actually find quite scary. Or something you think you can't or shouldn't do. Or simply because it is unfamiliar to you because you've never done it before. But letting go of what felt familiar gave me the space to experience who and what I really am. Because what was familiar to me had all been taught by my environment. The environment where I learnt to live and survive, but where I did not feel at ease.

And the more detached I became from it, the easier it was. Because speaking uncensored from my heart comes very easily to me. I love doing that and when I'm writing or typing my fingers can hardly keep up. I have always been quite impulsive, have taken decisions easily and acted on them immediately. That wasn't always convenient, but it was easy to do. And to me it felt right. So it didn't need much prompting. And the more I dared to let go of the old and familiar patterns the more I felt encouraged to do so. I felt it and responded to it. Letting go of the old patterns had only advantages. Nothing to lose and still I felt scared. But that too is programming. When I follow my heart I am no longer afraid. The voices that scare me then are other people's, not mine. My own quiet voice reassures me, the more I let go, the more I follow my heart.

Dreaming for the whole tribe

A long time ago, I did Integral Psychology training and we did a weekend of dream interpretation in a very large group. Just under 100 people. Hans Korteweg[1] told us then, "You can only share your dream if you are sure it benefits the whole tribe." And he also said something along the lines of: "If you feel your heart beating then you know it's your turn." Whereupon I raised my hand and told him my dream. And in that large circle, we worked out the whole dream. A deeply moving experience in many ways. Not just for me, but for the whole tribe. A very empowering experience which taught me to trust what naturally sprouted from me when I was given permission to let my inner voice speak.

I have always remembered this and have acted accordingly. And as a result, I have had the courage to take small steps as well as big leaps.

---

1) dekorteweg.nl

For the last 30 years, whenever I speak to a room full of people or I deliver a workshop or training, I think about this. I am at my best when I work this way.

Of course I prepare beforehand, but the less rigid the preparation is, the better the story appears. It's never the same, it's different every time. And even I don't know exactly what is going to happen.

The key is to trust myself and stand by my story or my message. And in the meantime, not be concerned with the audience grasping it all straight away. It didn't always go well. It went wrong when I thought about it too much. If I didn't feel at ease with the audience. And afterwards it became clear to me that it just wasn't the right place for me. A bit embarrassing sometimes, but I got over that. And I got better at choosing myself instead of waiting to be chosen. I learnt to follow my heart more. That's how I ended up in that beautiful place in Pelion, where the praying mantis jumped on my hat and where I started to do my retreats.

And that is also how I find more and more trust in myself and in my abilities. By doing things I've never done before. And going to places I've never been before. When I first saw Robert Smith on the internet, I knew immediately that I wanted to know more. I saw him working with an addict and I wanted to do that too. Soon after, I was in Budapest to do the first training with Robert.

And a year and a half after that, I was in Hawaii with him, volunteering with his team of, as he called it, "12 best practitioners from around the world" at a long-term drug treatment centre.

Every person was voluntarily admitted to the facility[1] and most had criminal records. There were no fences and anyone was able to leave at any time. Sometimes that did indeed happen.

The individual with the addiction really had to want to change, no matter how difficult it was. Every day we helped whoever wanted to be helped. And a year later, I was there again. This experience greatly strengthened my confidence in my mastery of this work.

The first step was the most important one. I knew clearly what I wanted and flew to Honolulu. It was daunting and it cost me a few thousand Euros, but I knew I was where I wanted to be. It was good for me and for the tribe. I may have been called a loner, but when it's right I am totally in my element in a group.

It feels right

To feel my heartbeat and hear my inner quiet voice, I had to let go of a lot of stuff. Over and over again. All sorts of things came at me from all sides. Old voices, old beliefs, fear, disbelief, mistrust and learnt shit. Untruths in my system. Limiting beliefs I had been taught and more of that unnecessary crap.

---

1) habilitat.com/

A couple of years ago, I underwent a heart scan because on a few occasions, I had experienced a few cramps. The scan was okay and so was everything else.

I take care of my heart by letting go. And I listen to my heart more and better. And when my heart beats loudly, I know it's my turn. That's why I wrote this book. And it feels the right thing to do.

10. How about you?

I asked the four specialists at the end of the talk, "What is your advice to someone with trauma?" And this is what they've told me:

Ina Oostrom:

The relationship between a therapist and the person asking for help is very important. Placebo research shows that a placebo given by a very empathetic doctor works better than a good medicine given by a non-fine doctor. The human mind decides what to do with medicine. So I think one of the most important things is that the relationship feels right.

It is a misconception that good therapy has to take years. If you find the root of the problem quickly you can pull it out and then someone can recover very quickly. And with hypnotherapy, we can just get to the root cause of the trauma. It's about finding the root cause first, which allows you to stand very differently in yourself. Not cutting off the branch and the leaves, because they will grow back in no time. But pulling everything out, root by root.

Martijn Groenendal:

Negative emotional reactions and the negative consequences of trauma are all things you have to let go of. That is easier said than done. It takes a decision by someone to grow to that point. And to develop themselves. To seek help. Living with trauma is a huge price to pay. Then you have a huge leakage of your life energy.

What happened is over. But what do I do with it now? You can't change pasts. People are not trained in that. The advantage of suffering from your old trauma is that you have to look for help. Or when you have a goal, but you can't move forward because your trauma gets in the way.

It inspires you to look for help. Either the ideal is very big or the pain is very big, making you realise you have to do something to change.

You can put in a new story. You can even start by putting the new story in, making the old story lose its power. By starting to write down for yourself who you are without the trauma. That requires work for the person who wants to resolve it. And you have to apply processes, which most people are not familiar with.

Self-hypnosis is a repetitive process. It's a lame metaphor, but if you want to go from overweight to a muscular body in the gym, that's a nice development. Nice goal and the desire is there too. But there is then a continuous process of discipline and accountability.

It can be compared to growing muscles. They are a different kind of muscle, but there is work involved. And that is the opposite of taking a pill or smoking marihuana or drinking. The next day, your problem is there again. You have to have the willingness and take responsibility. The law of cause and effect. If I stop doing that which is good for me, I notice it after a few months. Then I get punished for it. Then I can get frustrated or don't want to, that doesn't help me. Until I just go back to doing what is needed.

Hypnosis is a huge enrichment to add to your daily rituals. Which can make life super or less painful. If you get good at it, you can get super good results. You can realise your desires and let go of the past. We haven't really learnt to choose a particular goal for ourselves. Just act normal and you'll be crazy enough. You have children. You have a car. What more do you want? Your goals are often an afterthought or even an unnecessary luxury. It all seems fine, but inside it doesn't feel that way. But we can and do choose our thoughts.

Rob de Groof:

The most important thing is not to keep walking around with it. Talk about it with people around you. And if that doesn't work or if you don't feel up to it, talk about it with a therapist. And definitely don't bottle it up, because then you will reinforce those negative feelings. That's the downside of our subconscious, which is quite the lazy part of our brain. And if that gets a negative slant and you keep throwing negativity at it, it's going to backfire. And then if you come up with something positive or want to feel a bit better then of course that's not going to work anymore. So do hypnotherapy or NLP or Faster EFT. Don't wait too long with it. The longer you put it away the more it affects you.

People should realise that hypnosis is not a panacea. It is sometimes expected to be. Whatever session or technique you choose. People need to learn to let go of high expectations and trust the subconscious mind to make the changes. And learn to not conclude too fast that this 'also does not work'. You also have to give hypnosis and yourself a chance and some time.

Kim Jewell:

Most people who have been traumatised by something will work at any cost to avoid going near that trauma again. What they don't realise is that thinking, because it's in that survival technique, is going to recreate more patterns, which produces more evidence. And which makes that trauma even bigger. I call it the filing cabinet of thinking. When we take off the emotional charge and we change the perception, the thinking is not going to pull up those old triggers. And then they are free to move on in their lives. So I would say to anyone who feels resistance: it's your alarm bell. If we pay attention to this and we change it, then we free ourselves and our lives. Then something new can enter us. Once we take the underlying emotions offline, there is a domino effect on all the other parts of the pattern. Those were in place, but now all fall over.

When you feel your heartbeat

When you feel your heartbeat you know it's your turn.

At the end of each chapter, I asked you a few questions. You have this book in your hands for a reason. Something drew you in when you saw the back cover or the introduction or maybe just when you saw the title.

Now that you've read the book, you may be inspired to start letting go of what you don't want to drag around anymore yourself. In any case, answering the 24 questions will give you a lot of insight. You can use the text and the video (via the bonus 5 link) to learn to tap. In case you want to do more than that? Take a look at my website, which has more videos or check the information of the 4 specialists.

Here you have the 24 questions again:

1. Is there something from the past you would like to let go of?

2. Have you ever been forced to let go. For example, because of death or divorce?

3. Or do you (perhaps) let go too quickly, after which you start having doubts?

4. Do you remember activities you could get absorbed in (as a child)? And do you still do that sometimes?

5. How often do you take the time to do what you really enjoy?

6. Do you still do anything entirely for yourself, which is not necessarily useful.

7. What is the story of my parents?

8. What is the story of other educators who had a big role in my life?

9. What becomes possible when I let go of those old stories?

10. How do you feel about yourself now, about how you look and what you do?

11. Where do those beliefs come from?

12. And is it really true what you have come to believe about yourself?

13. Are you willing to make some effort?

14. Do you dare to stand alone?

15. Can you renounce what you are used to or attached to?

16. Do you have cravings you would like to give in to?

17. Have you perhaps found compensation so you don't have to give in?

18. What is stopping you from letting go and giving in?

19. What emotions keep surfacing for you?

20. Do you find you can't manage to push these emotions away, no matter how much you drink or snack or work?

21. Imagine what your life would be like if you could cut out these emotions. Write it down.

22. What new experience is already waiting inside you?

23. You've probably experienced laughing in a crying fit. Or vice versa. Have you had such an experience?

24. Our emotions are close together and replace each other with ease. Have you ever experienced that?

A reminder of what I wrote in the introduction:

- ➢ Buy yourself a lovely notebook and pen. I can assure you that writing by hand has a better outcome than typing on a screen.
- ➢ No need for technology when you embark on this type of self-examination.
- ➢ Sit down in a comfortable place, where you won't be disturbed.
- ➢ Take 5 minutes to answer each question. Set a timer and do nothing else in those few minutes.
- ➢ Write as fast as you can and do not pay attention to punctuation or spelling. As long as you understand what it says, it's good enough.

Take time for it and take time for yourself. You can read all the books in the world, but when you don't put it into practice there will be no results. We all have to do the work ourselves, because nobody else can do it for you. You are unique and special and worthwhile. You deserve to live rather than survive. Know that you are worthy enough to do it.

And now YOU learn to tap and let it go

Take all the answers you wrote in response to the 24 questions.

Take a sheet of paper and draw a line in the middle of it. Put a - (minus sign) on the left and a + (plus sign) on the right. Or print the pdf you see at bonus 5 on the website.

In the left column, write everything you want to get rid of. And in the right-hand column, write what you do or want more, better. Re-read everything you wrote down earlier in response to the 24 questions above. And put it in the right columns. When there's more coming up, write it down and put it on your lists as well. This is not a one time action. You can do it as often as needed or as often as you like.

You speak to your own subconscious mind during this process and on that level everything you say is immediately accepted and understood, even if your conscious mind doesn't agree. No explanation is needed there. Keep it simple! Put it in one or a few words in the column. Make sure the list is FULL. At least 24 points. Research has shown that you only get to the really important stuff above 20.

A simple form for your list is in the back of this book, but it's much easier when you use the bonus material, print it out and work with it. That's why I made it and I put them on this website. See the video and attachments to know how and where to tap according to my method.

The way I use tapping is based on Faster EFT. In Clinical EFT, there are more acupuncture points, which are tapped. But you can skip those, which makes it work faster. And the big advantage of the way I use and teach it to you now, is that you use only your own words. And that resonates directly with you. Unlike text that is not relevant to you, like in many general scripts.

Choose the right words when you talk to yourself

After you have watched the instructions and know how to tap, you start listing the list point by point. You tap and say, "Whatever it is, I release it and let it go."

"Whatever it is" you can replace with a point on your list, for example, "All pain, all anger, all sadness, I release it and let it go." Or: "All the pain I have experienced in my relationship with ... , I now release it and let it go. It is time to let it go. It's safe to let it go."

When you take the words on the right you just say them, confirm and affirm all the good on your list. I am ... I have ... etc.

Work through your list like this. You can do the left side first and than the right side of your list. Or you alternate constantly.

Personally, I like to alternate using the left and right columns. And I also alternate tapping and stroking. You can't feel bad and good at the same time, so it collapses at the moment you do that.

174

What you can also alternate is the way you tap and where you tap. See the 2nd page of variations in the attachment. So many things are possible. Try it out.

For example, I tap on the left side of my face and say, "All the fatigue I am feeling right now I am releasing and letting it go. It's time to let go. And then I stroke the right side of my face and say, "I am deeply relaxed and enjoying the energy flowing through my body." Remember that everything you give attention to grows.

Energy always goes to where you focus your attention. So start paying attention to what you say, even if no one hears it. Because YOU do hear it. Throughout the day, that very self-talk is what defines you. So do you choose to undersell yourself? Or do you give yourself a good mental boost, encouragement, a sweet compliment or a pat on the back every so often? Pay attention to it. You can influence your day, your mood and the outcome of what you achieve in a day by paying attention to how you talk to yourself. And adjust that if necessary. The only one who hears everything you say and think is you! Thoughts are creations. They do not come easily. And when they do ... you can choose what to do with them. These lists are just an example. Take time for it and don't stop till you drop. ;-) It's about creating your life, you know. That's worth some time!

"Energy flows where attention goes."

Kahuna wisdom

BONUS 5

The 5th bonus is copied and you can find them as the attachments at the end of this book. There's also the video, where I explain and show you how to tap in many different ways for all kinds of reasons. See bonus 5[1]

---

[1] tapandletgo.com/bonus5

Epilogue

A big thank you to everyone, who made it possible for me to live this life and write this book. To my beloved family for giving me lots of stories to write and always keep going, no matter what.

I am grateful for my wonderful daughter Litô. Her constant support, creative contribution, thoughtful comments and wonderful sense of humour. Having her precious presence around is always a bliss.

Thanks to all the brave clients, who gave me their trust and for being open minded for receiving and enjoying the amazing and surprising results. Through the success of the sessions and by recording and looking back, I am getting better at understanding what we do and how it works. This reinforced the urge to write this book.

Some people took time to read, comment on and improve this book. I thank Lynn Hogendoorn, Martina Eicher, Karen Sjouke, Maarten Hamburger, Sarah Kotzamani and my dear friend Frank van Rooij.

All the expertise, experience and advice, shared with me by Ina Oostrom[1], Kim Jewell[2], Martijn Groenendal[3] and Rob de Groof[4], have enriched this book and me. And for this I sincerely thank these four passionate specialists. Delightful!

---

1) hypnosementor.nl & hypnosecommunity.nl
2) kimjewell.com.au
3) hypnotherapieacademienederland.nl
4) hypnosecentrum.be & europeanhypnosisacademy.com

Dominique Pickerill[1] has done the translation in English. Words are powerful and it's a challenge to try to express yourself in any other language about any subject dear to the heart. I'm so grateful Dominique did this amazing job and she really put her heart in it. It's not only about translating words. It is sensitive material and Dominique knows about the work we both do. It was a huge learning experience for me and fun to communicate about the book in all ways possible. We phoned, we texted, we skyped, we chatted and we whatsapped (yes, I know these are not proper verbs). Finally we had a good reason to meet again and joined some days together in Wales at the seaside, a plan we had for years.

Special thanks as well to Marina Malthouse[2] and Sasha Papadin[3] for recording the audiobook. People have told me that reading my stories often feels like they hear me talking and I like that, but in the end it's not important that it sounds like me. I've asked Marina for the audiobook, because what really matters is that people like to listen to the storyteller and I love to listen to the sound of her voice. For that matter it's important that Marina is having a good time reading and recording the book. So I was very happy to hear that Sasha, who's a musician, sound professional and good company, offered to work with us.

---

1) calmandfocus.co.uk
2) bhma.org/what-holistic-healthcare-means-to-me-dr-marina-malthouse
3) thisistrac.com

CJ Kale[1] is a wonderful photographer from the big island in Hawaii, who's work I greatly admire. Nature is fascinating and I can't get enough of watching the wild sea, deserted deserts and volcanoes. The people in Hawaii have to know how to let go, because they live on a volcano. Like CJ, who generously told me to use his photos for my book covers. So lucky and grateful I am for having such amazing people around me.

Last but not least I thank Elia Stephanidou[2]. Not because she helped me with the book, she did not know until recently that I had written it. I thank her because she was the right woman in the right place at the right time to lift me up and carry me home, take care of me and help me get rid of a trauma that had shaped my life since childhood. I lay on her couch, I slept in her bed, I peed on her floor and I cried the best cry ever. She helped me walk, made me laugh, gave me good food and strong painkillers with Cypriot coffee. The best! She even organised a dance party with our Cypriot crew when the doctor wouldn't let me stand on my feet for 6 weeks. She gave me a perfect session, drove me in a wheelchair to my plane, saved my life and made me a better human being.

1) lavalightgalleries.com
2) eliastefanidou.com

Book tips

"I am a hypnotist and so is your mother."

Richard Bandler

Here's a very small selection from my favourite books. When you understand what hypnosis is, you start to recognise it and see how promoters, priests and politicians use it too. Just pay attention when you watch television (tell-a-vision). Do not only listen to the content of the words chosen, but also pay attention to hand gestures, facial expressions and tone of voice.

Hypnosis is a powerful tool to use for the benefit of others and yourself. Anyone can do it. So can you!

Not all the books below has hypnosis as a subject, officially, but they are all about the power of the mind, I call it Your Inner Power, YIP.

Breaking the habit of being yourself
Joe Diepenze (2012)

Feeling is the secret
Author: Neville Goddard (1944)

From Stress to Success
The Secrets to Fast, Permanent Life Change with FEFT (2012)
Author: Kim Jewell

Hypnose de sleutel tot eigen kracht (in Dutch)
(Hypnosis the key to self power) Explanations, application and

scientific directions for adults and children (2017)

Author: Ina Oostrom

Hypnotherapy

Author: Dave Elman (1984)

Kill the hypnotist

Based on true life events of hypnotist Tom Silver (2021)

Author: Suzanne Silver

Mentalism for Hypnotherapists

Learn How to Use Mentalism as a Gateway to Hypnosis (2019)

Author: Rob de Groof en anderen

Monsters and Magical Sticks

There is No Such Thing as Hypnosis (2005)

Authors: Stephen Heller, Terry Steele and Robert Anton Wilson

My voice goes with you . . .

Therapeutic stories of Milton H. Erickson (2011)

Collected and commented by: Sidney Rosen

Operatie met hypnose (in Dutch)

(Surgery with hypnosis) Removing a tumour from the breast with hypnosis as natural anaesthesia, a personal report (2018)

Authors: Ina Oostrom en Mirjam Borsboom

Stop It!

The Art of Simple Hypnosis (2022)

Author: Rob de Groof

Reframing – NLP

Authors: Richard Bandler and John Grinder (1985)

The Amazing Life of Ormond McGill

Dean of American Hypnotists (2005)

Author: Ormond McGill

The Deep Trance Training Manual Hypnotic Skills

Author: Igor Ledochowski (2003)

The Red Book

Author: Carl Jung

Unconscious learning

Author: Milton H. Erickson (1992)

And all other books from these authors. My list could be as big as this book. I read and write every day. It's a lovely way to improve your mind.

About the author

Jeannette van Uffelen (1959) is a behavioural specialist and stress release expert with 40 years of experience in helping children and adults from all over the world, who has earned her spurs in the field of healing from emotional pain, anxiety and trauma. She worked in crisis care, child protection, vocational training and addiction services and started her own practice as an entrepreneur in 2003. She studied Applied Integral Psychology and is an advanced certified professional and trainer in ICT, social skills, empowerment, reprogramming, DNA healing, hypnosis and EFT tapping.

She offers her services in her 'Your Inner Power' (YIP) workshops, retreats[1] and private sessions[2] in several European countries and online.

She teaches people how to reprogram their mind and apply relaxation techniques on themselves and release the effects of trauma. She makes this practically simple to apply. She wrote her first book 'Undercurrent' in 2007. About TAP & LET GO she says:

"No method is sacred, neither is mine. And there is no one-size-fits-all. But there are insights and simple techniques, which I want to share with as many people as possible. That is why I have written this book."

---

1) yipretreats.com/
2) jeannettevanuffelen.com/

Attachments

And now it is your turn
TAP & LET it all GO

While reading or listening to the book, did you take the time to ask yourself the questions for each chapter and **write down your answers?**

No? Not a disaster. The book seems to be a nice read and before you know it you are in the next chapter. But NOW you can still sit down for this.

I know some readers have scrolled straight on to chapters 6 and 7 to read my drama-trauma story. And some have left the rest for what it was.

Others have read the whole book but not done anything with the assignments assuming they don't need it, don't have time for it or will do it later.

I often did that too, but I am learning. Nowadays, I tend to listen to a book 10 times, while walking, shopping, washing the dishes, vacuuming, et cetera. And strangely enough, I hear something new every time.

There are some books that I have listened to at least 50 times and one even 100 times. That one now sounds like a trusted friend I can't get enough of.

You are so smart that you sit down for it NOW. Maybe you already have the answers. If not, on the next page you will see all the questions listed again.

Write down the answers. Grab a notepad or a few sheets and go wild! Let the ink flow. The smoother the better and don't limit yourself. Only you need to be able to follow it, no one else. So no censorship.

Print out the sheet with the table or take a sheet of paper and draw some lines, as on the following sheet. Put a - (minus sign) on the left and a + (plus sign) on the right.

In the left column, write everything you want to get rid of. And in the right-hand column, write what you do or want more of. Re-read everything you wrote down earlier in response to the questions above. And put it in the right columns.

During this process, you talk to your own **subconscious** system and everything you say is immediately understood there. No explanation is needed there. Keep it simple! Put it in one or a few words in the column. Make sure the sheet is FULL. Minimum of 24 points. **It seems you only get to the really important stuff after 20.**

The 24 'How about you?' questions

1. Is there something from the past you would like to let go of?
2. Have you ever been forced to let go. For example, because of death or divorce?
3. Or do you (perhaps) let go too quickly, after which you start having doubts?

4. Do you remember activities you could get absorbed in (as a child)? And do you still do that sometimes?
5. How often do you take the time to do what you really enjoy?
6. Do you still do anything entirely for yourself, which is not necessarily useful.

7. What is the story of my parents?
8. What is the story of other educators who had a big role in my life?
9. What becomes possible when I let go of those old stories?

10. How do you feel about yourself now, about how you look and what you do?
11. Where do those beliefs come from?
12. And is it really true what you have come to believe about yourself?

13. Are you willing to make some effort?
14. Do you dare to stand alone?
15. Can you renounce what you are used to or attached to?

16. Do you have cravings you would like to give in to?
17. Have you perhaps found compensation so you don't have to give in?
18. What is stopping you from letting go and giving in?

19. What emotions keep surfacing for you?
20. Do you find you can't manage to push them away, no matter how much you drink or snack or work?
21. Imagine what your life would be like if you could cut out these emotions.

22. What new experience is already waiting inside you?
23. You've probably experienced laughing in a crying fit. Or vice versa. Have you had such an experience?
24. Our emotions are close together and replace each other with ease. Have you ever experienced that?

-	+

188

Example

-	+
pain	I am healthy
I do not have any ...	there's always enough
powerless	I love myself
anger	I am pretty
frustration	my body is perfect
fear	I trust myself
the pain from the past	the past is over
the hurt child	I am an adult now
the memory of ...	I am free to choose what I want
what I have done	I forgive myself
what I regret	I choose consciously
what I have survived	I am flexible
negative selftalk	I am strong
not good enough	I grow and develope every day

189

TAP & LET it all GO

BASIC tapping instructions

Tap above your eye and say:
"I release and let it go".

Tap beside your eye and say:
"It is time to let it go".

Tap under your eye
and say: "Whatever
it is, I now let it go".

Tap on your collar bone and say:
"It is safe to let it go".

Take hold of your wrist and say:
"PEACE".

190

TAP & LET it all GO

VARIANTS tapping instructions

Variant 1: Do the sequence at the other side of your face.

Variant 2: Tap with both hands on both sides of your face.

Variant 3: Stroke with both hands on your face and collar bone.

Variant 4: Stroke with one hand on one side of your face and collar bone.

Variant 5: Put on your biggest smile and say "PEACE".

Do you want to TAP & LET GO some more?

And how was it for you doing this by yourself?
I'd loved to hear it. juffelen@pm.me.

Tapping 'in the heat of the moment' has benefited me a lot. When I started applying this, my life became calmer, easier and more fun. I am happier and do more what I want and what is right for me. Much of what I thought was impossible for me I have now realised and experienced. I was fed up with the city and have been living and walking for months now in all kinds of wonderful places in the countryside of Germany, Slovenia, Greece, England, Italy, Wales and the Netherlands.

In terms of work, I am also doing more of what I really want to do. I am writing more and I have become my own publisher, so I can start sharing much more of my knowledge and experience.

And a deep desire I have also realised and it tastes like more. I really love retreats because they take you away from your daily life for a while. In retreats I do not focus on DOING differently but on BEING differently. Because I believe the solution is there. It is here inside me. But that is sometimes very hard to find, because I am programmed with all kinds of beliefs and I'm full of habits. Do you recognise that? Then check out my offer of our magical Your Inner Power retreats in beautiful locations.

And of course, individual sessions have also helped me release trauma. You could have read or listened to this in the book. In addition to the retreats, I continue to give individual sessions, because it's beautiful work and this enables true transformation.

Let us hear from you!

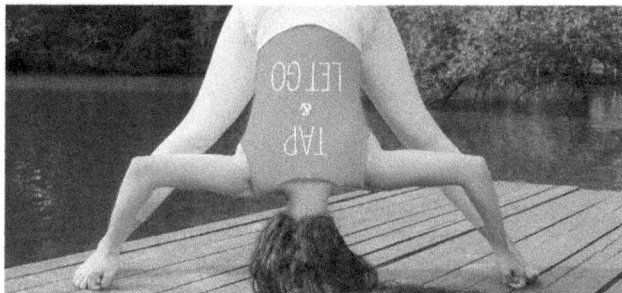

copyright © Jeannette van Uffelen - tapandletgo.com

Attachment: tapping instructions to go

Tap above your eye and say:
"I release and let it go".

Tap beside your eye and
say: "It is time to let it go".

Tap under your eye
and say:"Whatever it
is, I now let it go".

Tap on your collar bone and say:
"It is safe to let it go".

Embrace with one hand your
other wrist and say: "PEACE".

Cut this page out and take it with you. Practice tapping. It is so easy.
Get used to tapping in the heat of the moment and feel how fast it works.

Milton Keynes UK
Ingram Content Group UK Ltd.
UKHW030104080824
446615UK00004B/256

9 798224 199471